Frontiers in Natural Product Chemistry

(Volume 6)

Edited by

Atta-ur-Rahman, *FRS*
Kings College
University of Cambridge
Cambridge
UK

Frontiers in Natural Product Chemistry

Volume # 6.

Editor: Atta-ur-Rahman, *FRS*

ISBN (Online): 978-981-14-4846-1

ISBN (Print): 978-981-14-4844-7

ISBN (Paperback): 978-981-14-4845-4

©2020, Bentham Books imprint.

Published by Bentham Science Publishers Pte. Ltd. Singapore. All Rights Reserved.

need for a court order if at any point you breach any terms of this License Agreement. In no event will any delay or failure by Bentham Science Publishers in enforcing your compliance with this License Agreement constitute a waiver of any of its rights.

3. You acknowledge that you have read this License Agreement, and agree to be bound by its terms and conditions. To the extent that any other terms and conditions presented on any website of Bentham Science Publishers conflict with, or are inconsistent with, the terms and conditions set out in this License Agreement, you acknowledge that the terms and conditions set out in this License Agreement shall prevail.

Bentham Science Publishers Pte. Ltd.
80 Robinson Road #02-00
Singapore 068898
Singapore
Email: subscriptions@benthamscience.net

CONTENTS

i

PREFACE

Frontiers in Natural Product Chemistry presents recent advances in the chemistry and biochemistry of naturally occurring compounds. It covers a range of topics, including important researches on natural substances. The book is a valuable resource for pharmaceutical scientists and postgraduate students seeking updated and critically important information on bioactive natural products.

The five chapters in this volume are written by eminent authorities in the field. Chapter 1 presents an overview of different ways of production to obtain bioactive peptides from different underutilized plant sources, including from food, brewing and bioethanol industries. Chapter 2 deals with the research on the design of new fluoroquinolones with improved features by molecular hybridization technique. Chapter 3 deals with a pathway for waxy rice starch hydrolysis by a compressed hot water process. Chapter 4 deals with the most important marketed plant alkaloidal drugs and their metabolites. Chapter 5 provides an insight into the molecular basis of preventive and therapeutic effects of natural bioactive substances against cancer diseases.

I hope that the readers will find these reviews valuable and thought-provoking so that they may trigger further research in the quest for new and novel therapies against various diseases. I am grateful for the timely efforts made by the editorial personnel, especially Mr. Mahmood Alam (Director Publications), and Mrs. Salma Sarfaraz (Senior Manager Publications) at Bentham Science Publishers.

Atta-ur-Rahman, *FRS*
Kings College
University of Cambridge
Cambridge
UK

List of Contributors

Ahmet Demirbas	Karadeniz Technical University, Department of Chemistry, Trabzon, Turkey
Antonio Lama-Muñoz	Department of Chemical, Environmental and Materials Engineering, University of Jaén, Campus Las Lagunillas, Jaén, Spain
Bikash Debnath	Natural cum Advance Synthetic Lab, Department of Pharmacy, Tripura University (A Central University), Suryamaninagar, Tripura, India
Dev Bukhsh Singh	Department of Biotechnology, Institute of Biosciences and Biotechnology, Chhatrapati Shahu Ji Maharaj University, Kanpur-208024, India
Eulogio Castro	Department of Chemical, Environmental and Materials Engineering, University of Jaén, Campus Las Lagunillas, Jaén, Spain Center for Advanced Studies in Earth Sciences, Energy and Environment, University of Jaén, Jaén, Spain
Francisco Espínola	Department of Chemical, Environmental and Materials Engineering, University of Jaén, Campus Las Lagunillas, Jaén, Spain Center for Advanced Studies in Earth Sciences, Energy and Environment, University of Jaén, Jaén, Spain
Kuntal Manna	Natural cum Advance Synthetic Lab, Department of Pharmacy, Tripura University (A Central University), Suryamaninagar, Tripura, India
Manish Kumar Gupta	Department of Biotechnology, Faculty of Science, Veer Bahadur Singh Purvanchal University, Jaunpur-222003, Uttar Pradesh, India
Manuel Moya	Department of Chemical, Environmental and Materials Engineering, University of Jaén, Campus Las Lagunillas, Jaén, Spain Center for Advanced Studies in Earth Sciences, Energy and Environment, University of Jaén, Jaén, Spain
Minerva Cristina García Vargas	Department of Chemical, Environmental and Materials Engineering, University of Jaén, Campus Las Lagunillas, Jaén, Spain Department of Industrial Engineering of Tecnológico Nacional de México/Instituto Tecnológico de Zitácuaro, Av. Tecnológico, 186 CP 61534, Zitácuaro, Michoacán, Mexico
Neslihan Demirbas	Karadeniz Technical University, Department of Chemistry, Trabzon, Turkey
N. Shimizu	Research Faculty of Agriculture, Hokkaido University, Hokkaido, 060-8589, Japan Field Science Center for Northern Biosphere, Hokkaido University, Hokkaido, 060-8589, Japan
Rajesh Kumar Pathak	School of Agricultural Biotechnology, Punjab Agricultural University, Ludhiana-141004, Punjab, India
T. Ushiyama	Graduate School of Agriculture, Hokkaido University, Hokkaido, 060-8589, Japan
Waikhom Somraj Singh	Natural cum Advance Synthetic Lab, Department of Pharmacy, Tripura University (A Central University), Suryamaninagar, Tripura, India

CHAPTER 1

Plant Protein Hydrolyzates from Underutilized Agricultural and Agroindustrial Sources: Production, Characterization and Bioactive Properties

María del Mar Contreras[1,*], Minerva Cristina García Vargas[1,2], Antonio Lama-Muñoz[1], Francisco Espínola[1,3], Manuel Moya[1,3] and Eulogio Castro[1,3]

[1] Department of Chemical, Environmental and Materials Engineering, University of Jaén, Campus Las Lagunillas, 23071 Jaén, Spain

[2] Department of Industrial Engineering of Tecnológico Nacional de México/Instituto Tecnológico de Zitácuaro, Av. Tecnológico, 186 CP 61534, Zitácuaro, Michoacán, Mexico

[3] Center for Advanced Studies in Earth Sciences, Energy and Environment, University of Jaén, Spain

Abstract: Today, there is a growing interest in the valorization of agricultural and agroindustrial waste/byproducts, including through obtaining bioactive compounds. Besides the use of plant proteins in animal nutrition, obtaining protein hydrolyzates could give an added value, improving digestibility and exerting functional properties by the generation of bioactive peptides. Bioactive peptides encrypted in plant proteins are latent until released and activated by proteolysis. Generally, to obtain bioactive peptides, enzymatic hydrolysis by peptidases is the most common way, with or without previous solubilization and purification steps of the intact protein. This hydrolysis step can be combined with physical and chemical treatments not only to improve the recovery but also to enhance the bioactivity. Therefore, our chapter presents an overview of different ways of production to obtain bioactive peptides from different underutilized plant sources, including from food, brewing and bioethanol industries. In order to characterize bioactive peptides, the application of conventional methods and more sophisticated methods based on mass spectrometry is also described. Moreover, recent literature on the bioactive properties of those plant peptides and current challenges associated with safety issues are discussed.

Keywords: ACE-inhibitor, Antioxidant, Antihypertensive, Antidiabetic, Byproduct, Bioactive peptide, Carbohydrolase, Hydrolysis, Mass spectrometry,

---

*** Corresponding author María del Mar Contreras:** Department of Chemical, Environmental and Materials Engineering, University of Jaén, Campus Las Lagunillas, 23071 Jaén, Spain;
E-mails: mcgamez@ujaen.es;mmcontreras@ugr.es;mar.contreras.gamez@gmail.com

Microwave assisted extraction, Peptidomics, Peptidase, Protein, Sustainability, Valorization.

INTRODUCTION

Considering the population factor and the state of natural resources, the need to look for more efficient agroindustry processes is recognized due to demographic growth and the current unsustainable practices. The global population is growing, while our standard of living is increasing; thereby, we have to face environmental challenges. In this sense, it is expected that the world's population increases by 2 billion people in the next 30 years and could reach around 11 billion in the next century. Based on this prognosis, it is not difficult to understand why the United Nation's second priority objective for the present century is to "End hunger, achieve food security and improve nutrition and promote sustainable agriculture" [1]. Moreover, the current agricultural and food practices also threaten the health of people and the planet: i) 70% of worldwide water use is required by agriculture; ii) it generates huge levels of pollution and waste; iii) risks associated with poor diets are one of the leading causes of death; iv) a double burden of malnutrition exists since millions of people are either eating not enough or eating the wrong types of food. In 2017, this led to one in eight adults (*i.e.* more than 672 million people) in the world to be obese [2] and forecasts suggest high levels of obesity on the future population [3]. In particular, increased demand for animal-based protein is expected to have a negative environmental impact, generating greenhouse gas emissions, requiring more water and more land [4]. Thereby, plant proteins could be an alternative but sustainable practices are required.

Therefore, against this background, the goal is how to meet the growing global demand for food, including protein and healthy foods, to improve income and employment in rural areas and, at the same time, reduce the environmental impact. This puts pressure on the world's resources to provide not only more but also different types of food, including more sustainable production of existing sources of protein as well as alternative sources for human consumption [4]. This requires us to move from an oil based economy towards a more sustainable circular bioeconomy model, producing more food and bio-based products from renewable resources, including agricultural and agroindustrial byproducts [5]. In this line, the biorefinery concept has emerged as a sustainable processing of biomass into a portfolio of marketable food and feed ingredients, bio-based products (chemicals, materials, proteins, bioactive compounds, *etc.*) and energy (fuels, power, heat) [6 - 8].

When thinking about these resources, plant compounds are usually put forward as their most probable source [9]. This includes macro (cellulose, hemicelluloses, pectins, starch, lignin, proteins, minerals, *etc.*) and microcomponents (*e.g.* phytochemicals). In particular, plant byproducts are underutilized sources of proteins and, most of the time are addressed to animal nutrition, but the ruminal degradability of proteins is not high. Nonetheless, proteins can be beneficial not only in terms of nutrition but also from a functional point of view through the generation of bioactive peptides. This means that the breakdown of peptide bonds by enzymatic hydrolysis increases the solubility, digestibility, and functional properties of the precursor proteins and byproduct [6]. Bioactive peptides are known for their high tissue affinity, specificity and efficiency in promoting health [10]. Therefore, apart from the use of plant proteins in animal nutrition, obtaining protein hydrolyzates could give an added value in a biorefinery context, with improved digestibility and exerting functional properties through the generation of bioactive peptides. This could also lead to the formulation of functional ingredients that are in line with the increased consumer awareness towards functional foods, nutraceuticals and personalized diets; the driving force of the functional food and nutraceutical market [10]. Moreover, there is a growing interest in the food industry and among consumers in reducing the use of synthetic additives in food preservation and opting instead for natural ones [11]. All this together connects with the concept of bioeconomy since it can promote a new way to diversify plant byproducts.

Generally, hydrolysis by peptidases, with or without a previous protein extraction step, is the most common way to obtain bioactive peptides with a wide range of biological properties, *e.g.* antidiabetic, antihypertensive, antimicrobial, antioxidant, and anticancer properties [12 - 15], but also autolysis and application of microbial suspensions (whole cells) have been applied [13, 16]. Enzymatic hydrolysis can be combined with physical treatments and alkaline extraction not only to improve the recovery but also to enhance the bioactivity [6, 17]. In this context, this book chapter presents an overview of the different ways of production to obtain bioactive protein hydrolyzates from different underutilized plant sources. These sources include byproducts from the cereals industry (wheat germ protein, broken rice by-product), oil industry (olive, and rapeseed/canola byproducts), fruit and vegetable industries (*e.g.* fruits seeds, potato byproducts, cauliflower leaves), and brewing industry (brewer's spent grain). Some techniques applied to characterize the hydrolyzates and the peptides are also covered. Moreover, the biological properties of the hydrolyzates have been revised, and the sequence of some bioactive peptides is shown. Finally, some safety issues are also discussed.

WAYS OF PRODUCING PLANT PROTEIN HYDROLYZATES FROM AGRICULTURAL AND AGROINDUSTRIAL SOURCES

There are several ways to produce bioactive hydrolyzates and peptides from natural sources: gastrointestinal digestion, fermentation, enzymatic hydrolysis, and genetic recombination [18]. This can be performed through the use of endogenous and exogenous microorganisms and proteolytic enzymes [19]. In general, microbial alkaline proteases can be produced under culture conditions. *Bacillus* produces extracellular proteases during post-exponential and stationary phases. Their proteolytic system contributes to the liberation of bioactive peptides [13]. Examples of commercial *Bacillus* enzymes are Alcalase, Neutrase, Esperase, Thermolysin, *etc.* Another enzyme employed is Flavourzyme from *Aspergillus oryzae.* Among them, Alcalase is widely used, probably explained by its high hydrolytic activity [20].

In the case of the agricultural and agroindustrial byproducts/waste the most common way to obtain bioactive peptides is through enzymatic hydrolysis using exogenous proteolytic enzymes. A wide variety of smaller peptides are generated *via* hydrolysis, depending on the byproduct, enzyme specificity and hydrolysis time [6, 15]. The hydrolysis conditions and enzyme type also affect the degree of hydrolysis and bioactivity due to the generation of peptides with a different amino acid sequences and length [15, 20 - 23]. Sometimes, gastrointestinal digestion enzymes are applied after enzymatic hydrolysis and ultrafiltration since it can enhance the bioactivity of the hydrolyzates, but it depends on the hydrolysis and ultrafiltration conditions [20, 24, 25]. Other times simulated gastrointestinal conditions are tested to assess the maintenance of the bioactivity as it was *in vivo* or to find the active peptides [26, 27]. Other studies have revealed that autolysis and fermentation were also plausible ways to recover bioactive peptides from potato and rice byproducts respectively [13, 16]. While the results of autolysis were improved after hydrolysis using *Bacillus* peptidases [16], the application of suspensions of strains of *Bacillus subtilis*, *Bacillus pumilus* and *Bacillus licheniformis,* which were isolated from different food sources, were successful in generating antioxidant peptides from rice compared to enzymes [13].

Concerning enzymatic hydrolysis, which is the most common way, it can be classified secondarily according to when the hydrolysis step with peptidases is performed; *i.e.* obtaining proteins and hydrolysis separately (sequential extraction and hydrolysis) (SeEH) or simultaneously (simultaneous extraction and hydrolysis) (SiEH). The proteolysis step can be performed with enzymes and preparations of enzymes from different origins, *i.e.* microbial enzymes, plant enzymes, and animal enzymes (Table **1**). Their effectiveness depends on the method and conditions applied as well as the primary sequence of the proteins.

While some enzymes, such as those with subtilisin activity (*e.g.* preparations like Alcalase) have broad specificity, others cannot work well if they are not favored by the primary sequence of the protein [6]. Other enzyme preparations consist of a mixture of peptidases and other enzymes types: *e.g.* Flavourzyme from *Aspergillus oryzae* contains two aminopeptidases, two dipeptidyl peptidases, three endopeptidases, and one α-amylase [28]; pancreatin and Corolase PP contain mainly trypsin and chymotrypsin from animal pancreas, but also other enzymes [29, 30]. The methods applied to generate bioactive peptides from agricultural and agroindustrial byproducts are detailed in the sections below. However, this book chapter is focused on the application of these methods in agricultural and agroindustrial byproducts/waste rather than on the description of the mechanism behind solubilization or their application in other matrices, which has been described in other reviews [6, 19, 31, 32].

Moreover, it should be considered that as a first screening, experimental assays with peptidases can be performed by changing one factor at a time and keeping others fixed, whereas response surface methodology can take several factors into account at the same time (as an example, see [14]). This includes screening tools like factorial designs with a reduced number of assays [33]. Moreover, *in silico* tools can be applied to choose those enzymes that are able to cut proteins theoretically on the basis of the primary sequence and the extent: *e.g.* the Peptide-Cutter of ExPASy (https://web.expasy.org/peptide_cutter/; accessed on 17 January 2018) and EnzymePredictor (http://bioware.ucd.ie/~compass/biowar eweb/; accessed on 17 January 2018) [6, 34]. These tools include enzymes like: Arg-C proteinase, caspases, chymotrypsin, pepsin, thermolysin, thrombin, staphylococcal peptidase I, trypsin, enterokinase, glutamyl endopeptidase, proline-endopeptidase, *etc.*

Table 1. Enzymes reported for the hydrolysis of proteins from agricultural and agroindustrial byproducts/waste. The information about the enzymes was retrieved from the BRENDA database (https://www.brenda-enzymes.org/; accessed 22 January 2020).

Enzyme	EC Number	Reaction
Pepsin A	3.4.23.1	Preferential cleavage: hydrophobic, preferably aromatic, residues in P1 and P1' positions
Chymosin	3.4.23.4	Broad specificity similar to that of pepsin A
Chymotrypsin[a]	3.4.21.1	Preferential cleavage: Tyr╪, Trp╪, Phe╪, Leu╪
Trypsin[a]	3.4.21.4	Preferential cleavage: Arg╪, Lys╪
Subtilisin[b]	3.4.21.62	Broad specificity for peptide bonds, stereoselective hydrolysis of amino esters
Thermolysin[c]	3.4.24.27	Preferential cleavage: ╪Leu > ╪Phe
Bacillolysin[d]	3.4.24.28	Similar to that of thermolysin

(Table 1) cont.....

Enzyme	EC Number	Reaction
Papain[e]	3.4.22.2	Broad specificity for peptide bonds, preference for amino acids with a large hydrophobic side chain at the P2 position

[a]Pancreatin and Corolase PP enzyme preparations contain these enzyme activities, among others.
[b]Alcalase 2.5L, Protex 6L, Esperase contain mainly serine endopeptidase (mainly subtilisin A).
[c]Protex 14L contains thermolysin activity.
[d]Neutrase contains bacillolysin activity.
[e]Corolase L10 and Promod 144MG contain papain activity.

Sequential Extraction and Hydrolysis (SeEH)

If SeEH is selected (Table **2**), the main driving force of the protein extraction process should be taken into account, *i.e.* chemical, biochemical, physical (including, mechanical) and physical-chemical, but a combination of them is also common [6, 31]. In any case, the steps basically consist of: conditioning, extraction/solubilization of the plant protein, concentration (purification or isolation)/drying and hydrolysis by peptidases [6, 35].

Extraction by Chemical Methods and Hydrolysis

In general, these methods normally require a previous grinding of the byproduct, the use of homogenization, shearing and/or thermal treatments to enhance the protein solubilization. Then, the use of neutral solutions, acid solutions, alkaline solutions, organic solvents, salt solutions, surfactants and reducing agents have been applied to recover proteins from plants [31, 36 - 39]. It also enables the fractionation of cereals proteins, recovering most proteins (98%); *e.g.* using consecutively, sodium chloride (albumins and globulins), ethanol (prolamins), acetic acid (acid-soluble glutelins), and sodium hydroxide (residual proteins) aqueous solutions [40]. In general, these extraction methods have been reviewed by several publications [31, 37, 38] and, in particular, by our recent review [6], which was focused on plant proteins from agri-food byproducts/residues. Concerning the obtention of protein hydrolyzates, Table **2** exemplifies methods to recover proteins from these sources and the subsequent hydrolysis conditions. In addition, Table **3** shows some of the characteristics of the protein products obtained before and after hydrolysis.

Among these methods, alkaline solutions are widely used as a generally recognized as safe (GRAS) solvent for the extraction of proteins from cereals and seed storage proteins [39], as well as agri-food byproducts [6 - 8, 17, 41]. In particular, Connolly *et al.* (2013) [39] reported the use of aqueous-alkaline conditions to recover proteins from pale brewer's spent grain, which was subjected to protein hydrolysis in subsequent works to generate bioactive peptides [25, 35]. The protein recovery was up to 59%, and hordeins, glutelins and low molecular weight peptides (41% < 10 kDa) were recovered. This value is lower than those reported by Qin *et al.* [36] (85-95%), who tested acid conditions, either in a single step or two-step treatments (after an alkaline extraction step) [36]. However, proteins were probably obtained in the form of amino acids owing to the strong conditions used. Moreover, alkaline extraction has been also applied to recover proteins from broken rice, canola, palm kernel and walnut cakes after oil extraction [15, 23, 25, 35, 39, 49, 50] with different extraction conditions: pH from 8 to 13.0, from room temperature to 50 °C, and extraction time from 45 to 180 min, with and without NaCl. Then, hydrolysis with alkaline peptidases, mainly Alcalase, has enabled the obtention of antimicrobial, antioxidants, anticancer and ACE inhibitory (ACEi) peptides, while anticancer peptides were obtained using papain (Table **2**).

Other studies have applied buffered extraction at neutral pH, using detergents like sodium dodecyl sulfate (SDS) and reducing agents with, for example, dithiothreitol, to recover proteins from olive seeds [20, 51]. For food applications, these conditions should be adapted to food grade conditions.

Extraction by Enzymatic Methods and Hydrolysis

Adding enzymes during protein extraction increased protein recovery and yield compared to control experiments without enzymes [17]. This included the use of carbohydrolases and peptidases to assist in protein extraction. The first type assists by degrading the cell wall components, while the second type assists through proteolysis. However, only studies focused in the use of carbohydrolases are discussed in this section, while the use of peptidases is the objective of the section below about direct SiEH methods.

Most of studies involving carbohydrolases have been performed at neutral to acidic pH conditions and applied thermal treatments (45-55 °C) for 2-6 h. They also applied one main enzyme activity like α-amylase and pectinase, but most of them used enzyme preparations and their mixture (Table **2**). Examples of the use of carbohydrolases are found applied to brewer's spent grain [12, 24, 35] and rice bran [48], Moreover, a treatment with carbohydrolases was performed before alkaline extraction to recover proteins from defatted wheat germ and dehulled rapeseed [42, 47]; in the first case it enabled a better recovery of oil. All these studies used consecutively the action of peptidases to promote protein extraction and hydrolysis (Table **2**). This can be also considered SiEH methods if the use of carbohydrolases is more like a pretreatment of the biomass. In this sense, it seems that the use of peptidases is more effective in extracting biomass than only the use of carbohydrolases [17]. For example, impure food-grade amylase containing protease was more effective than Celluclast and Viscozyme in protein extraction from heat-stabilized defatted rice bran [43], with protein recoveries up to 58.4%. Nonetheless, some studies on brewer's spent grain have applied a pretreatment with carbohydrolases (*e.g.* cellobiohydrolases, endoglucanase, xylanase and β-glucosidase or cellulase, xylanase (endo-1,4-), and β-glucanase (endo-1,3(4)-)) before the use of microbial peptidases [35, 44]. In particular, the results of Rommi and coworkers showed that the use of a steam explosion pretreatment and a carbohydrolase cocktail had no influence on the theoretical yield of protein solubilization. Interestingly, the technical protein yield was improved using one of the treatments before the use of peptidases, which was probably related to a reduced water binding capacity. In another work, the combination of the carbohydrolases cocktails (Depol 740 and Econase CE) and peptidases from *Bacillus* (Alcalase and Promod 439) was applied to release sugars and to recover the proteinaceous material from brewer's spent grain [45, 46]. It enabled to extract more than 80% of the proteinaceous material and up to 39% of the total carbohydrates were solubilized; either in simultaneous or sequential treatments, while the order of the enzymes did not affect the results.

Moreover, not only the protein recovery can vary but also the bioactivity (*e.g.* antioxidant, DPP-IV inhibitory, and immunomodulatory activity) if hydrolysis by carbohydrolases is applied as a pretreatment before proteolysis compared to alkaline extracted proteins [35]. These authors concluded that direct hydrolysis with carbohydrolases and peptidases, for example, using the combination Alcalase and Flavourzyme, is a less resource-intensive process to obtain bioactive peptides compared to the alkaline extraction of proteins before proteolysis. In another work, Zhang and coworkers [47] applied pectinase, cellulase, and β-glucanase followed by sequential treatments consisting of alkaline extraction and peptidase treatment (Alcalase 2.4L) to both produce a protein hydrolyzate product and demulsify the oil from rapeseed.

Extraction by Physical/Physical-chemical Methods and Hydrolysis

This includes the use of homogenization [31], high pressure-assisted extraction and new trends such as ultrasound-, microwave- and electro-assisted extraction (including pulsed electric field and high voltage electric discharge) [6]. Generally, the shearing effect provided by the homogenization step is not enough to promote effective breakdown and extraction of proteins, and it should be combined with chemical agents [25, 31, 39]. Conversely, ultrasound can favor the recovery of proteins, but also modify their properties, as Li *et al.* [52] reported on bell pepper seeds. Ultrasound has been applied at different operational conditions to assist buffered and alkaline extraction of proteins from fruit (olive, plum and peach) seeds and cauliflower leaves [14, 20, 21, 26, 51]. These studies found that the subsequent hydrolysis by microbial enzymes generated antioxidant and ACEi peptides (Table **2**). Another study evaluated the effects of ultrasound applied before and during proteolysis of defatted wheat germ protein. Its result revealed that an ultrasonic treatment of 20 kHz (900-1800 W) during proteolysis facilitated the hydrolysis with Alcalase, whereas ultrasonic pretreatment promoted the release of ACE inhibitor peptides during enzymatic hydrolysis. Moreover, the study by Ding *et al.* [53] has shown that an ultrasonic pretreatment, using different frequencies and working modes, improved the degree of hydrolysis of grape seed protein and the bioactivity of the hydrolyzates.

Table 2. Summary of conditions applied to obtain plant protein hydrolyzates from agricultural and agroindustrial byproducts in sequential extraction and hydrolysis (SeEH). The protein content (PC), recovery (PR) and/or protein yield (PY) of the initial and final products are indicated, as well as the bioactivity of the hydrolyzates.

Byproduct	Summary of the Conditions Applied		PC/PR Initial (%)	PC/PR Final (%)	Bioactivity[a]	Ref.
Cereals						
Defatted wheat germ protein	Hydrolysis with α-amylase (E/S, 1/100, w/w) (55 °C, 2 h), alkaline extraction (pH 9.5, adding 1 M NaOH, 30 min). Acid precipitation (pH 4.0)	Hydrolysis with Alcalase 2.4L FG (pH 8.0, maintained using 1 M NaOH, 50 °C, 210 min); E/S (up to 0.4:100, w/w). This was assisted by ultrasound (40 W)			ACEi activity (IC_{50} from around 0.5 to 2.5 mg/mL)	[42]
Defatted wheat germ protein	Hydrolysis with α-amylase (E/S, 1/100, w/w) (55 °C, 2 h), alkaline extraction (pH 9.5, adding 1 M NaOH, 30 min). Acid precipitation (pH 4.0)	Pretreatment with ultrasound (0-1800 W) and then hydrolysis with Alcalase 2.4L FG (pH 8.0, maintained using 1 M NaOH, 50 °C); E/S (up to 0.4:100, w/w)			ACEi activity (IC_{50}, 0.26-0.42 mg/mL)	[42]

(Table 2) cont.....

Byproduct	Summary of the Conditions Applied		PC/PR Initial (%)	PC/PR Final (%)	Bioactivity[a]	Ref.
Pale and black brewers' spent grain	Oven-drying, ground, and aqueous (with shearing) and alkaline extraction (110 mM NaOH, s/l 1:20) (twice) (50 °C, 1 h). Acid precipitation (pH 3.8)	Preconditioning step and hydrolysis with Alcalase 2.4L, Corolase PP, or Flavourzyme (1%, w/v; w/w) (pH 7, maintained using 2 M NaOH, 50 °C, 240 min)	PC: 18-46 PR: 15-59		ACEi activity (IC$_{50}$, 0.32-0.69 mg/mL). The best values were obtained by Alcalase	[25,35,39]
Brewers' spent grain	Extraction of phenolic compounds and carbohydrate hydrolysis	Hydrolysis with alkaline protease + Flavourzyme (AF) and neutral protease + Flavourzyme (PF) (5:100, w/w) (pH 7-9.5, maintained with 2 M NaOH, 50-55 °C, 2 h each enzyme). Simulated gastrointestinal digestion (AFD and PFD, respectively) and ultrafiltration (1 kDa)		PC: 33-48	All samples showed thrombin inhibitory activity. Peptides <1 kDa from hydrolyzates and digested samples also delayed thrombin and thromboplastin time respect to the control (~63%) ACEi activity (IC$_{50}$ values, 1.4-5.2 mg/mL of protein)	[24]
Brewer's spent grain	S/l 10% in water for shearing and then hydrolysis with carbohydrolases, (Shearzyme and UltraFlo) (75 μL g^{-1} byproduct each one) (pH 5.0, 50 °C, 4 h) to recover the solid fraction	Hydrolysis with Alcalase (2%, g^{-1} protein) (pH 9, 50 °C, 2 h) and then with Flavourzyme (1%/g protein) (pH 7.0, 50 °C, 2 h)		PC: 45	Hypotensive effects *in vivo*, ACE (76.7%) and DPP-IV (83.9%) inhibitory activity, as well as antioxidant activity (ORAC, 614.1 μmol TE/g protein)	[12]

(Table 2) cont.....

Byproduct	Summary of the Conditions Applied		PC/PR Initial (%)	PC/PR Final (%)	Bioactivity[a]	Ref.
Steam-exploded brewer's spent grain	Hydrolysis with carbohydrolases (12 mg protein/g): 60% Cel7A (cellobiohydrolase I), 10% Cel6A (cellobiohydrolase II), 15% Cel5A (endoglucanase), 12% Xyl10 (xylanase) and 3% Cel3A (β-glucosidase). S:l 20% (6 h, 50 °C) and filtration to recover the solid fraction	Hydrolysis with an alkaline protease (Biotouch Roc 250L) (10 mg total protein/g substrate) (pH 10, 0.2 M Na-carbonate, 50 °C, 5 h and maintenance of the pH). Acidification to pH 2.5 and centrifugation to recover the supernatant	PC: 31	PC: 39		[44]
Brewer's spent grain	S/l 10% in water for shearing and then hydrolysis with carbohydrolases, (Shearzyme and UltraFlo) (75 µL g^{-1} byproduct each one) (pH 5, 50 °C, 4 h) to recover the solid fraction	Hydrolysis with Alcalase (2%, w/w protein) (pH 9, 50 °C, 2 h) and then with Flavourzyme (1%, w/w protein) (pH 7.0, 50 °C, 2 h). Other combinations tested: Corolase PP and Flavourzyme, Prolyve 1000 and Protease P and Alcalase and Protease P		PC: 45 PR: 63	Inhibition of DPP-IV: 35.7-85.1%[b] Antioxidant activity: 238-287 (FRAP) and 1255-1128 (ORAC) µmol TE/g protein[b]	[35]

(Table 2) cont.....

Byproduct	Summary of the Conditions Applied		PC/PR Initial (%)	PC/PR Final (%)	Bioactivity[a]	Ref.
Brewer's spent grain	Aqueous (with shearing) and alkaline extraction (110 mM NaOH, s/l 1:20) (twice) (50 °C, 1 h). Acid precipitation	Hydrolysis with Alcalase (2%, g^{-1} protein) (pH 9, 50 °C, 2 h) and then with Flavourzyme (1%, g^{-1} protein) (pH 7.0, 50 °C, 2 h). Other combinations tested: Corolase PP and Flavourzyme, Prolyve 1000 and Protease P and Alcalase and Protease P	PC: 46	PR: 59	Inhibition of DPP-IV: 87.0-89.6%[b] Reduction in the release of interleukin-6 in lipopolysaccharide-stimulated RAW 264.7 cells (immunomodulatory activity): 58.3/48.0%[b] Antioxidant activity: 266-296 (FRAP) and 1207-1384 (ORAC) µmol TE/g protein[b]	[35]
Milled and defatted rice bran		Alcalase 2.4 L or Flavourzyme (pH 8.0, 50 °C); E/S (0.5%, w/w of protein). The supernatant was obtained after centrifugation		PC: 28-30 PR: 81-88		[22]
Heat-stabilized defatted rice bran	Dispersion in water, acidification (pH 3.5) and hydrolysis with pectinase (45 °C, 2.5 h)	Hydrolysis with Protease P (pH 7.0, 45 °C, 3.1 h); E/S (1200 units/g of byproduct)		PR: 80		[48]
Broken rice	Ground and alkaline extraction (pH 13, in NaOH solution), homogenization and maintenance at 50 °C, 3 h. Acid precipitation	Enzymatic hydrolysis by alkaline protease (pH 8.0, 50 °C, 1.5 h)-ultrafiltration (6 kDa)			Antioxidant activity by the DPPH radical scavenging activity	[49]
Oilseed/Industrial crops						

(Table 2) cont.....

Byproduct	Summary of the Conditions Applied		PC/PR Initial (%)	PC/PR Final (%)	Bioactivity[a]	Ref.
Defatted canola cake	Ground and alkaline extraction at pH 8 (0.05 M phosphate buffer containing 1 M NaCl, at room temperature, 60 min). Precipitation (ammonium sulfate)	Hydrolysis with Alcalase 2.4L (pH 8.5, 50 °C, 2.5-3 h); E/S (4%, w/w, on the basis of protein content)	PR: 50.		ACEi activity (IC_{50} 0.015 mg/mL)	[50]
Dehulled rapeseed seeds	Boiling, wet milling and hydrolysis with pectinase, cellulase, and β-glucanase (4:1:1, v/v/v) (pH 5, 48°C, 3 h)	Akaline extraction (pH 10) and hydrolysis with Alcalase 2.4L (pH 9.0, 60°C, 150 min); E/S (1.5%, v/w)		PR: 80–83		[47]
Palm kernel cake	Palm kernel expeller proteins were extracted using strong alkaline solution	Hydrolysis with Alcalase (pH 9.59, 50 °C); E/S (1%, w/w)			Antibacterial effect against most of the Gram-positive and spore-forming bacteria was at 70% DH	[23]
Oil-bearing fruits						
Olive stones	Ground and buffered extraction (denaturing and reducing conditions) of the seeds. Precipitation (cold acetone)	Hydrolysis with thermolysin (pH 8.0, 55°C, 2 h); E/S (0.05 g enzyme/g protein). Other enzymes tested: Alcalase 2.4 L FG, Neutrase 0.8 L, Flavourzyme 1000 L, and PTN 6.0 Saltfree*N			The highest ACEi activity was obtained by thermolysin (IC_{50} 0.029 mg/mL)	[20,51]

(Table 2) cont.....

Byproduct	Summary of the Conditions Applied		PC/PR Initial (%)	PC/PR Final (%)	Bioactivity[a]	Ref.
Olive stones	Ground and buffered extraction (denaturing and reducing conditions) of the seeds. Precipitation (cold acetone)	Hydrolysis with Alcalase 2.4L FG (pH 8.0, 55°C, 2 h); E/S (0.15 AU/g of protein). Other enzymes tested: Neutrase 0.8 L, Flavourzyme 1000 L, PTN 6.0 Saltfree*N and thermolysin			DPPH (68.6%) and ABTS radical scavenging capacity (72.0%), TEAC (363 μmol Trolox equivalents/g), hydroxyl radical scavenging activity (54.5%), and lipid peroxidation inhibition (91.2%)[b]	[20,51]
Walnut oil extraction residue	Alkaline extraction (pH 9.0, 45 °C for 1 h) with NaOH solution and acid precipitation	Hydrolysis with papain (pH 7.0, 60 °C, 3 h). Other enzymes used: alkaline protease, pepsin, trypsin and neutral protease			Best results with papain: inhibition towards cancer cells growth	[15]
Fruits						
Plum seeds	Ground, defatting, and buffered extraction (s:l 1:167) (denaturing and reducing conditions) with ultrasound (1 min, 30% amplitude). Precipitation (cold acetone)	Hydrolysis with Alcalase 2.4L (pH 8.5, 50 °C, 3 h); E/S (0.30 AU/g protein). Other enzymes tested: Thermolysin, Flavourzyme, and Protease P	PY: 39		• The highest antioxidant activity (TEAC) was found using Alcalase (460 μmol TE/g) • ACEi activity (IC_{50} < 0.5 mg/mL) (similar to that for Thermolysin)	[21]
Peach seeds	Ground, defatted, and buffered extraction (denaturing and reducing conditions) with ultrasound (5 min, 30% amplitude). Precipitation (cold acetone)	Hydrolysis with thermolisin (pH 8, 50 °C) and ultrafiltration (fraction <3 kDa). Other enzymes tested: Alcalase, Flavourzyme, and Protease P.	PY: 43		The highest activity was found using thermolysin (IC_{50}, 0.016 mg/mL)	[26]
Vegetables and tubers						

(Table 2) cont.....

Byproduct	Summary of the Conditions Applied		PC/PR Initial (%)	PC/PR Final (%)	Bioactivity[a]	Ref.
Cauliflower leaves	Homogenization and alkaline extraction (s:l 1:4, pH 11, 15 min) with ultrasound (175 W). Acid precipitation (pH 4)	Hydrolysis with Alcalase (pH 8, 55 °C, 4 h); E/S (5000 U/g)	PR: 53		• ACEi activity (IC$_{50}$, 0.14 mg/mL) • Promotion of the glucose consumption and enhanced the glycogen content in HepG2 cells	[14]

[a]*In vitro* angiotensin converting enzyme-inhibitory (ACEi) activity. *In vitro* antioxidant activity determined by TEAC (trolox equivalent antioxidant capacity), ORAC (oxygen radical absorbance capacity), FRAP (ferric reducing antioxidant power) assays, ABTS and DPPH radicals scavenging assays, hydroxyl radical scavenging activity, and lipid peroxidation inhibition. IC$_{50}$, concentration of inhibitor required to inhibit 50% of the ACE activity.
[b]Values for the enzyme providing the highest activity, whose hydrolysis conditions are described.
AAE, ascorbic acid equivalents; E/S, enzyme/substrate ratio; KA, kojic acid; S:l, solid-to-liquid ratio; Ref., reference; TE; Trolox equivalents.

Concerning the other extraction methods, further work is required since strong alkaline, acid and subcritical water extraction conditions can promote the hydrolysis of proteins [6, 7, 54], but probably not in a specific way as with using enzymes. In the case of microwave, it can favor the denaturation of proteins, but making them more accessible to peptidases [55], and thereby its combination can be interesting to favor the extraction of peptides [7]. For example, the use of microwave-assisted enzymatic hydrolysis (power of 300 W and treatment time of 2.5 min) improved the protein extraction from defatted corn gluten meal and shortened the hydrolysis time compared to a traditional water-bath enzymatic hydrolysis [56]. Alternatively, Ketnawa and Liceaga [57] have shown that the use of a microwave pretreatment followed by conventional enzymatic hydrolysis with Alcalase was the best treatment to produce antioxidant peptides and the lowest immunoreactivity from fish by-product.

Table 3. Some qualitative characteristics of plant protein products derived from agri-food byproducts and their hydrolyzates using sequential extraction and hydrolysis (SeEH) and simultaneous extraction and hydrolysis (SiEH). The extraction and hydrolysis methods are described in Tables 2 and 4.

Byproduct	Protein Product [a]	Hydrolyzate Product [a]	Analysis Method [a]	Ref.
SeEH – cereals				
Defatted wheat germ protein		All EAA, with or without ultrasound pretreatment	Amino acid analysis	[42]

(Table 3) cont.....

Byproduct	Protein Product [a]	Hydrolyzate Product [a]	Analysis Method [a]	Ref.
Brewer's spent grain	Hordeins, glutelins and low MW peptides. 45.5% of this fraction > 15 kDa. All EAA, except Trp	26.6-33.3% fraction >15 kDa[b]; high proportion of <2-5 kDa peptides[c]	SDS-PAGE; SE-LC; RP-LC; amino acid analysis	[25, 39, 35]
Brewer's spent grain		High proportion of <2 kDa and hydrophilic peptides[d]	SE-LC, RP-LC	[35]
Brewer's spent grain		All EEA. Four main peaks with MW >7000, 2100, ~500, and 120 Da. Fraction of 500 Da increased after digestion	SE-LC; amino acid analysis	[24]
Brewer's spent grain	MW > 10 kDa	High content of peptides with MW < 10 kDa, especially, < 1 kDa. More hydrophilic and hydrophobic peaks than unhydrolyzed proteins	SE-LC and RP	[12]
Broken rice		<200 Da-10 kDa before ultrafiltration; <200 Da-3 kDa after ultrafiltration	SE-LC	[49]
SeEH - oil-bearing fruits and oilseeds				
Defatted canola cake	Cruciferin, napin and other minor proteins	Peptides < 1.3 kDa (58.7%)	SDS-PAGE; SE-LC; DSC	[50]
Palm kernel cake	Some bands at 15 and 2.4 kDa. All EAA, except Trp and Met	Some bands at 15 and 2.4 kDa. All EAA, except Trp and Met	SDS-PAGE; SE-LC; amino acid analysis	[23]
Olive stones	Oleosin (50 kDa), storage proteins from the Solea II precursor (27 and 30 kDa), oleosin/storage proteins from the Solea I precursor (22 kDa). Hydrophobic and hydrophilic peaks	Peptides < 1kDa were identified after ultrafiltration (3 kDa)	SDS-PAGE; RP-LC; RP-LC-MS	[20, 51]
SeEH - fruits and vegetables				
Plum seeds	Bands below 75 kDa; intense bands between 37 and 50 kDa and lower than 25 kDa	Degree of hydrolysis around 62%; peptides characterized < 1 kDa	SDS-PAGE; RP-LC; RP-LC-MS	[21]
Peach seeds	Bands below 75 kDa; intense bands around 20 kDa	Degree of hydrolysis around 77%; peptides characterized < 1 kDa	SDS-PAGE; RP-LC; RP-LC-MS	[26]
Cauliflower leaves	All EAA		Amino acid analysis	[14]

(Table 3) cont.....

Byproduct	Protein Product [a]	Hydrolyzate Product [a]	Analysis Method [a]	Ref.
SiEH – cereals				
Rice protein byproduct		Intense bands around 50 kDa or with lower MW, which depend on the enzyme. Also presence of peptides with MW lower than 1.5 kDa from several proteins[c]	SDS-PAGE; SE-LC; RP-LC-MS	[13]
Rice protein byproduct		Intense bands between 75 and 15 kDa. Also presence of peptides with MW lower than 1.5 kDa, mainly from glutelin[f]	SDS-PAGE; SE-LC; RP-LC-MS	[13]
Liquid by-products from the processing of broken rice		Intense bands < 12 kDa	SDS-PAGE	[62]

[a] DSC, differential scaning calorimetry; EEA, essential amino acids; MW, molecular weight; RP-LC; reversed phase liquid chromatography; RP-LC-MS, reversed phase liquid chromatography-mass spectrometry; SDS-PAGE, sodium dodecyl sulfate-polyacrylamide gel electrophoresis; SE-LC, size-exclusion liquid chromatography; Trp, tryptophan; Met, methionine.
[b] Conditions reported in [25].
[c] Conditions reported in [35] using an aqueous-alkaline treatment.
[d] Conditions reported in [35] using a carbohydrolase treatment.
[e] Conditions reported in [13] using peptidases.
[f] Conditions reported in [13] using *Bacillus* suspensions.

Simultaneous Extraction and Hydrolysis (SiEH)

The use of peptidases during the solubilization of proteins has two main advantages, *i.e.* thermal-alkaline requirements can be reduced and, at the same time, proteolysis is accomplished [6]. This is also important from an industrial processing perspective as the scaling up of SiSH methods can be envisaged in a more amenable way. From an environmental perspective, it requires lower concentrations of alkali and significantly reduced energy inputs as the SiSH methods do not require the first concentration/drying steps [35]. Although the intact proteins cannot be obtained and so their primary functionality is lost, the hydrolysis products maintain the nutritional characteristics as being a source of amino acids. The use of peptidases during extraction may increase protein solubility and hydrolyzed proteins are attractive feed/food supplements because they have decreased antigenicity and are easier to digest [58]. Moreover, the proteolysis can also be modulated/optimized to generate bioactive peptides, as described below (Table **4**). Another interesting feature is that the use of peptidases can help to separate peptides from lignin after centrifugation as suggested Rommi *et al.* [44]. It is desirable, for example, when lignin is also solubilized since it can co-precipitate with proteins *via* acid precipitation, which reduces protein enrichment and thereby the protein content of the final product [7]. Accordingly, this strategy could be considered for lignin-rich byproducts.

In this context, SiEH methods have been applied to defatted distiller's grain, rice protein byproduct, sesame, and potato byproducts; mainly, using microbial enzymes at neutral and alkaline pH (up to pH 8) and with thermal treatments up to 55 °C. Anti-tyrosinase, anti-inflammatory, antioxidant and ACEi activities have been reported for some of these hydrolyzates (Table **4**). In particular, the extraction of zeins from distiller's grain has been shown to be difficult and with low recoveries using organic solvents alone or mixed with water. However, the use of Protex 6L (an alkaline serine-endopeptidase, subtilisin type) during extraction increased the recovery values to around 90% of the protein in both milled and unmilled defatted distiller grain. This was similar to that for alkaline-ethanol extraction (45% ethanol and 55% 1 M NaOH), although the latter was only effective for milled but not unmilled byproduct [58].

The use of physical methods, as a pretreatment of the biomass or to assist the proteolysis, is also plausible, but further work is required. For example, the use of an extrusion pretreatment has been applied to recover proteins through aqueous extraction by hydrolysis with peptidases [59, 60]. Moreover, Wang *et al.* [61] applied dual-frequency ultrasound and an immobilized enzymolysis treatment with Alcalase to promote the hydrolysis of isolated rapeseed protein. Comparing with no sonication, the degree of hydrolysis was increased at the optimal conditions (*i.e.*, an incubation time, power density and solid-liquid ratio of 76 min, 57 W/L and 5.3 g/L, respectively), as well as the yield of protein (64.61%), peptides (40.88%) and total sugar (23.60%).

Table 4. Summary of conditions applied to obtain plant protein hydrolyzates from agricultural and agroindustrial byproducts in simultaneous extraction and hydrolysis (SiEH) methods and other direct methods. The protein content (PC), recovery (PR) and/or protein yield (PY) of the initial and final products are indicated, as well as the bioactivity of the hydrolyzates.

Byproduct	Summary of the Conditions Applied	PC/PR Initial (%)	PC/PR Final (%)	Bioactivity[a]	Ref.
Cereals					
Defatted distiller's grain	S:l (w/v) 10:150 in deionized water and hydrolysis with Protex 6L (1:10, v/w) at pH 8.0 (maintained using 2 M NaOH) (50 °C, 2 h)		PR: 90		[58]
Rice protein byproduct	Homogenization and hydrolysis with Alcalase 2.4 L (pH 8, 55 °C, 2 h), Neutrase 0.8 L (pH 6.5 and 40 °C, 2 h) and Flavourzyme (pH 6, 60 °C, 2 h)			Antioxidant actitivity by ABTS; Neutrase hydrolyzate was the most active (≈ 150 µg AAE/mL)	[13]

(Table 4) cont.....

Byproduct	Summary of the Conditions Applied	PC/PR Initial (%)	PC/PR Final (%)	Bioactivity[a]	Ref.
Rice protein byproduct	Homogenization and hydrolysis with *Bacillus* strains (37 °C, 72 h) suspensions: *B. subtilis* (SV27, SV20I), *B. pumilus* (AGI) or *B. licheniformis* (AG2)			Antioxidant activity by ABTS; AG strains were the most active ($\approx$ 275 µg AAE/mL)	[13]
Liquid by-products from the processing of broken rice	Hydrolysis with Protamex (pH 8, 60°C, 2h); E/S (0.5 U/g of protein). Other enzymes tested: Alcalase 2.4L, Neutrase 0.8L, Flavourzyme 500L		PC: 4 g/L[b]	Anti-tyrosinase (0.3 g KA/L), anti-inflammatory (fold induction of NFκB with TNFα), antioxidant by ABTS assay (2.3 g AAE/L) and ACEi (18 mg/L) activities[b]	[62]
Oilseed/Industrial crops					
Dehulled sesame seeds	Ground, boiling and hydrolysis with Protex 7L (45 °C, 120 min). Other enzymes tested were: Alcalase 2.4L, Viscozyme L, Natuzyme, and Kemzyme		87		[63]
Vegetables and tubers					
Potato byproducts	Autolysis 5 h	PC: 12-49		• ACEi activity (IC_{50}, 0.042-0.070 mg peptide/mL) • Poor antioxidant activity by ABTS (30.8-45.3%)	[16]
Potato byproducts	Hydrolysis with Alcalase (pH 7, 55 °C, 5 h), Neutrase (pH 7, 50 °C, 5 h) or Esperase (pH 7, 55 °C, 5h); E/S ratio: 1/100 (w/w)	PC: 12-49		• ACEi activity (IC_{50}, 0.018-0.034 mg peptide/mL) • Antioxidant activity by ABTS (49-79%)	[16]

[a]IC_{50}, concentration of inhibitor required to inhibit 50% of the ACE activity. *In vitro* angiotensin converting enzyme-inhibitory (ACEi) activity. *In vitro* antioxidant activity determined by ABTS radical scavenging assay.
[b]Values for the enzyme providing the highest activity, whose hydrolysis conditions are described.
AAE, ascorbic acid equivalents; E/S, enzyme/substrate ratio; S:l, solid-to-liquid ratio; Ref., reference.

CHARACTERIZATION OF PROTEIN HYDROLYZATES

Conventional Methods

Crude protein of agricultural and agroindustrial byproducts is commonly measured by the Kjeldahl and Dumas (elemental analysis) methods. The factor applied to convert the nitrogen content to protein content depends on the plant source, according to WHO/FAO [64]. For example, 5.83 and 5.95 factors have been applied for rice byproducts [40, 42, 48], 5.9 for corn byproducts [58], and

6.25 is a general value applied for olive leaves, rapeseed meal, brewer's spent grain, sesame seeds, *etc* [7, 50, 65, 66]. These methods are also commonly applied to determine the protein content after hydrolysis [67]. Moreover, the colorimetric methods Bradford (Coomassie) (595 nm) [21, 68, 69] and Lowry (750 nm) [70, 71] has also been applied to estimate the protein content using bovine serum albumin as reference. Although these methods can be applied to measure the total content of peptides [71], Bradford is only recommended to for large peptides (> 3000 Da) [72]. The bicinconinic assay (BCA) (562 nm) [73] and the ortho-phthalaldehyde (OPA) ($\lambda_{excitation}$ 360 nm, $\lambda_{emission}$ 455 nm) [20, 21]; methods are other alternatives to measure the content of proteins and solubilized peptides, using BSA and glutathione as standards. All these methods can be run in microplates with some changes in the quantitative range but with advantages considering the amount of sample and reagent requirement and analysis time. Alternatively, Qubit® assay is another option that requires low amounts of soluble protein and peptides (https://www.thermofisher.com/order/catalog/product/Q33211#/Q33211; accessed 22 january, 2020). It requires a particular Qubit fluorometer and measures $\lambda_{excitation}$ 470 nm and $\lambda_{emission}$ 570 nm.

As an example, we found similar results of olive mill leaves proteins/peptides solubilized using mild and strong alkaline conditions [7], when the content was measured by Bradford assay (in terms of BSA) and elemental analysis (the factor applied to convert N into protein was 6.25) (Fig. **1A**). In this sense, this figure shows a very good relationship between Bradford assay and elemental analysis with a slope around 1 and an R^2 value of 0.901. Moreover, a good correlation was also found (R^2, 0.842) between Bradford and Qubit assays, but the results were slightly different (Fig. **1**).

Moreover, the amino acid analysis enables the measurement of the content of essential and non-essential amino acids. For that, there are several regulated methods. As an example, in the EU this determination is based on the use of an amino acid analyzer or high-performance liquid chromatography (HPLC) with ion exchange column, device for ninhydrin, post-column derivatization and photometric detector for the determination of all amino acids, while HPLC with fluorescence detection is suggested for tryptophan [74]. Another alternative could be RP-HPLC using UV detection at 262 and 338 nm as described [75]. The essential amino acid composition of some plant protein products obtained from agri-food by-products can be consulted in [6].

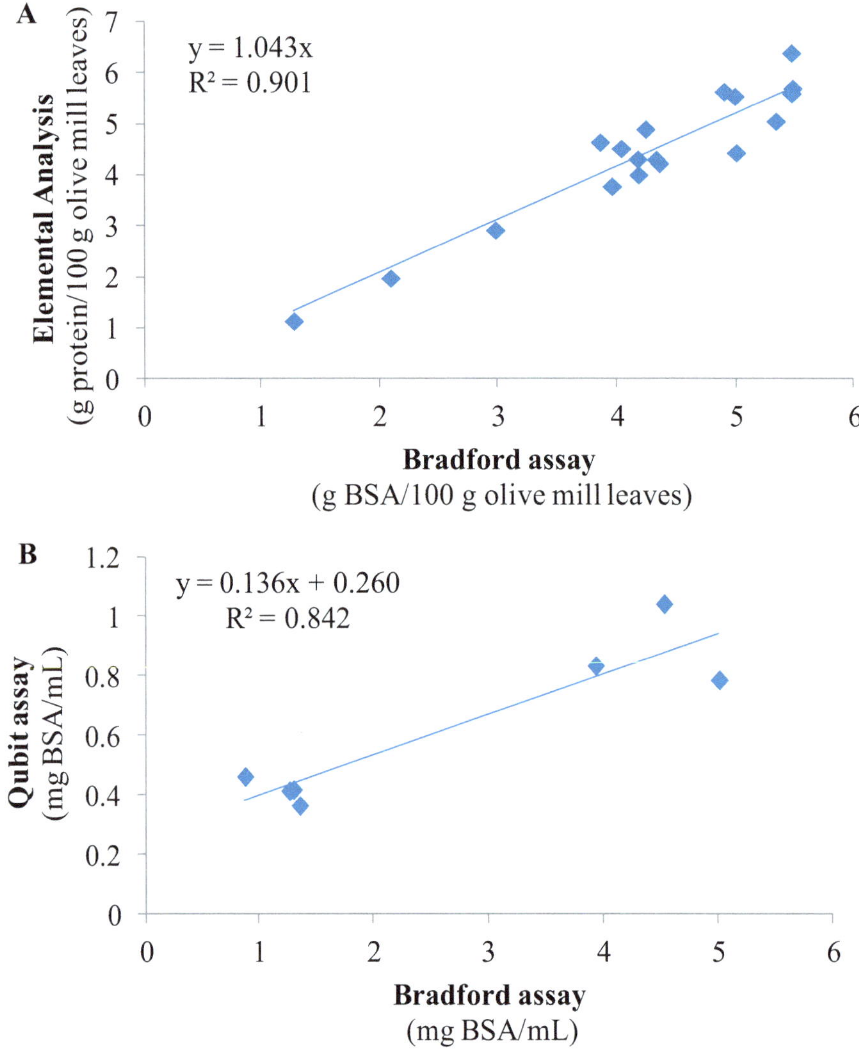

Fig. (1). Comparison of some assays to measure the content of soluble proteins/peptides.

Polyacrylamide Gel Electrophoresis

SDS-polyacrylamide gel electrophoresis (PAGE) has been used for size-based separations of byproduct proteins in either non-reducing conditions [54, 76] or reducing conditions using β-mercaptoethanol [39, 50, 76 - 78] and dithiothreitol [79], mainly. This can lead to different results. For example, Gerzhova *et al.* [76] used both reducing and non-reducing conditions to characterize an electro-activated protein concentrate of rapeseed. Under the reducing conditions two of the most distinctive bands of 16 and 45 kDa disappeared and smaller bands of 10, 12, and 22 kDa were generated, indicating the presence disulfide bonds between them. Other bands of 18, 27, and 32 kDa were observed, being not disaggregated by β-mercaptoethanol.

SDS-PAGE can be applied to detect proteins and alternatively hydrolysis products; *i.e.* plant proteins disappear or bands with lower molecular weight appear in the gel [7, 8, 44, 80] (Table **3**).

Another modality for protein characterization is two-dimensional electrophoresis, which is one of the simplest methods for this purpose and provides high resolution for the separation of complex protein mixtures [81]. This enables the separation of proteins in the function of both the isoelectric point (isoelectrofocusing) and the molecular weight (SDS-PAGE). As an example, this technique was used for the comprehensive characterization of olive seed storage proteins, with 1000 and 750 spots detected by silver staining in the endosperm and the cotyledon [82]. Obviously, the best applicability is found for soluble proteins but not for small peptides, due to poor resolution and potential elution with the front dye.

Capillary Electrophoresis

To a lesser extent, capillary electrophoresis (CE) has been used for the characterization of protein and peptides from plant byproducts. However, it has some advantages compared to other techniques in terms of small sample requirements and it is an environmentally-friendly technology. As an example, CE with a diode-array detector (DAD) at 214 nm was used to determine the protein profile of olive leaves. CE determination was performed under basic conditions using a background electrolyte (BGE) and an uncoated fused-silica capillary [83]. Alternatively, SDS-capillary gel electrophoresis (CGE) can be used for accurate protein quantification and molecular weight determination. The basic apparatus for CGE is identical to the aforementioned technique, but the separation media is a sieving matrix instead of the BGE solution [84]. This technique was used by the latter authors after enzyme-assisted extraction of the proteins and reduction by 2-mercaptoethanol. In this case, a total of 11 peaks were visualized, with molecular weights comprised between 20 and 300 kDa [85]. As far as we know, this technique has not been applied to characterize plant protein hydrolyzates from agri-food byproducts.

Liquid Chromatography

HPLC is the most widely used technique for both analytic and preparative separation of peptides, commonly coupled with conventional UV and fluorescence detectors as well as mass spectrometry (MS) [86]. A common modality to evaluate the molecular weight distribution of protein hydrolyzates is size exclusion (SE)-HPLC (Table **3**). For example, SE-HPLC at 214 nm has been used to determine peptides from above 0.2-14 kDa in brewer's spent grain hydrolyzates, which were correlated with the degree of hydrolysis [45, 71]. This technique was also used, for example, to analyze antioxidants hydrolyzates from a rice protein byproduct [13].

Another modality is a reversed phase (RP) mode that involves the use of non-polar stationary phases, such as C18, and enables the separation of proteins and peptides on the basis of hydrophobicity. It was applied to study the protein profile of plum and peach seed in stone residues using UV (210, 254, and 280 nm) and fluorescence detection ($\lambda_{excitation}$ 280 nm, $\lambda_{emission}$ 360 nm) [21, 26]. Moreover, due to advantages, such as analytical versatility, dynamic range and power for structure elucidation, MS has emerged as an important tool for protein and peptide characterization after chromatographic separation; especially when coupled to RP-HPLC [86]. Using MS and MS^n, peptides can be elucidated by *de Novo* sequencing and partial sequencing, also taking into account *m/z* measurements and information in databases. (Table **5**) shows examples of bioactive peptides, mainly sequenced by MS.

A new green modality of LC is nano-LC, which involves the use of capillary columns with internal diameters ranging between 10 and 100 µm. It provides a good compromise between sensitivity, loadability, and robustness. Moreover, it also presents an environmental advantage due to a lower solvent demand [89, 90]. RP-nano-LC-MS/MS was used to characterize tryptic hydrolyzates of olive seed proteins extracted with conventional extraction methods and combinatorial peptide ligand libraries (CPLLs) and subsequently separated by SDS-PAGE. CPLLs technology was used to reduce the protein dynamic concentration range. In fact, it enabled the identification of 61 proteins in olive seeds through the identification of peptides [91], a much higher number than using other methodologies. A similar strategy was used by this research team to characterize plum and peach seeds, enabling to find 141 and 96 proteins, respectively [92]. Recently, Dei Piu' *et al.* [13] used nano-LC-MS/MS to characterize 44 peptides, containing a number of amino acids ranging from 5 to 45 (450–4500 Da), in purified antioxidant fractions from rice protein byproduct hydrolyzates. Most of them were a mixture of peptides and derived from glutelins and other proteins.

Moreover, when the study is focused on the determination of a bioactivity, a fractionation step (*e.g.*, ultrafiltration and semi-preparative RP-HPLC, SE-HPLC or gel filtration chromatography) is added in order to identify the most active fractions and the active peptides [13, 20, 25, 26] (Table **5**). A representative example of this workflow was used by Connolly *et al.* [25], which sequenced 34 peptides in a bioactive fraction from brewer's spent grain hydrolyzed with Alcalase and digestive enzymes using RP-LC-MS/MS. Another option is to combine the analysis of the protein hydrolyzates by two HPLC modalities to increase the number of peptides identified, *e.g.* using reversed phase and hydrophilic interaction chromatography (HILIC) modes [26].

Table 5. Characterization of active peptides from plant protein hydrolyzates from agricultural and agroindustrial byproducts and bioactivity.

Byproduct	Peptide Sequence	Bioactivity[a]	Ref.
Brewer's spent grain	430 to 1070 Da peptides, rich in V, Y, and F	UF fraction (<1 kDa) with antithrombotic activity	[24]
	IPY	Antioxidant activity (0.37 μmol TE/μM peptide); DPP-IV inhibitory activity (IC_{50}, 352.15 μM)	[12]
	LPY	Antioxidant activity (0.34 μmol TE/μM peptide)	
	YPLP	Antioxidant activity (0.26 μmol TE/μM peptide)	
	IPLQP	ACEi activity (IC_{50}, 3.10 μM); DPP-IV inhibitory activity (IC_{50}, 105.45 μM)	
	LPLQP	ACEi activity (IC_{50}, 3.17 μM)	
	LPIA	DPP-IV inhibitory activity (IC_{50}, 45.07 μM)	
	IPVP	DPP-IV inhibitory activity (IC_{50}, 38.67 μM); ACEi activity (IC_{50}, 26.20 μM)	
	IVY	ACEi activity (IC_{50}, 80.4 μM)	[25]
	ILDL	ACEi activity (IC_{50}, 96.4 μM)	
	VHSP	ACEi activity (IC_{50}, 173 μM)	
	HHM*P	ACEi activity (IC_{50}, 226 μM)	
	ILLPGAQDGL	DPP-IV inhibitory activity (IC_{50}, 145.5 μM)	[87]
Corn α-zein	LQP	ACEi activity (IC_{50}, 1.9 μM)	[88]
	LLP	ACEi activity (IC_{50}, 57 μM)	
	LSP	ACEi activity (IC_{50}, 1.7 μM)	
	LAA	ACEi activity (IC_{50}, 13 μM)	
	FY	ACEi activity (IC_{50}, 25 μM)	
	LPP	ACEi activity (IC_{50}, 9.6 μM)	
	VHLPP	ACEi activity (IC_{50}, 18 μM)	
	VHLPPP	ACEi activity (IC_{50}, 200 μM)	

(Table 5) cont.....

By-product from rice starch industry	SYQDVYN	Identified in the most active antioxidant fractions, and presence of Y	[13]
	KSYQDVYN		
	KSYQDISVSA		
	YKSYQDVYN		
	QYKSYQDVYN		
	YPGLSN		
	FQQQYYPGLSN		
	KSYQDISVSA		
	KSYQDVYNVAESS		
	QYKSYQDVYN		
Walnut oil extraction residue	CTLEW	Selective inhibition towards cancer cells Caco-2, Hela and MCF-7 cells, which could be mediated by inducing apoptosis and autophagy; immunomodulatory activity	[15]
Olive seed	FDGEVK	Identified in an antioxidant fraction and presence of Y, F, M, and/or H	[20]
	VLSPPFTGE		
	MDGAP		
	APGAGVY		
	VVVVPH		
	MGSPT		
Olive seed	LTPTSN	Identified in an ACEi fraction	[20]
	LVVDGEGY		
	FDAVGVK		
	AFDAVGVK		
	VGVPGGV		
	LLPSY		
	ALMSPH		
	LFSGGES		
	LMSPH		
	LPAGA		
Peach seed	LYSPH	Showed common features of antihypertensive peptides	[26]
	LYTPH		
	HLLP		

[a]ACEi, angiotensin I-converting enzyme inhibitory activity; DPP-IV, dipeptidyl peptidase-IV; IC_{50}, concentration of inhibitor required to inhibit 50% of the ACE activity; Ref., reference.
*oxidised residue.

Matrix Assisted Laser Desorption Ionization (MALDI)-MS

MALDI-MS is based on the production of ions by laser bombardment of crystals containing a small amount of analyte (protein/peptide) dispersed in a large amount of matrix, which acts as a chromophore for the laser radiation, and an analysis by MS [86]. This technique has been used to characterize the aqueous fraction derived from the enzyme-assisted extraction peanuts kernels, which was applied to extract oil and protein. This study revealed the presence of peptides (107) with molecular masses from m/z 700 to 2369 Da [66].

Others

Circular dichroism and Fourier transform-infrared spectroscopy are useful to analyze the secondary structure changes of the proteins when subjected to extraction [93] and during hydrolysis [61]. For example, the former technique was used to evaluate the effects of ultrasound-assisted enzymolysis, which decreased the number of α-helix and a random coil of rapeseed protein by 10.7% and 4.5%, respectively, while β-chain increased by 2.4% [61]. Both techniques were also applied to evaluate changes in the secondary structure of rice residue protein when alkali was used at high pH values [93].

Moreover, fluorescence spectroscopy, recorded at $\lambda_{excitation}$ = 280 nm and $\lambda_{emission}$ from 290 to 500 nm, was applied to evaluate potential complexation between proteins/peptides and phenolic compounds from flaxseed [94]. In addition, surface hydrophobicity can be measured using fluorescence spectroscopy (390 nm and 400–650 nm as excitation and emission wavelengths) and the probe 1-anilino-8-naphthalene-sulfonate [42].

Structure characteristics can be also measured by scanning calorimetry. As an example, Wu and Muir [50] determined the thermal properties of canola protein, observing: an endothermic trough with two overlapping peaks at 84 and 102 °C, as well two parallel transition peaks attributed to cruciferin and [95, 96] napin proteins.

BIOACTIVE PROPERTIES OF PROTEIN HYDROLYZATES AND ACTIVE PEPTIDES

As commented before, hydrolysis is a way to obtain bioactive peptides from plant proteins with a wide range of biological properties as shown by recent reviews, including antioxidant properties [19, 97], antimicrobial activity [95, 96], quelating ability for dietary mineral bioavailability enhancement [98] and antihyperglycaemic (α-amylase inhibitory) activity [99]. In particular, (Tables **2** and **4**) show the bioactivity of plant protein hydrolyzates obtained from several agricultural and agroindustrial by-products. In addition, (Table **5**) details the bioactivity of particular peptides characterized in some of these hydrolyzates, as commented before.

Antioxidant Activity

Most studies have been focused on obtaining antioxidant plant peptides from agricultural and agroindustrial byproducts (Tables **2** and **4**). Antioxidant compounds exert their activity by two main mechanisms: hydrogen transfer and electron donation. The first method is commonly achieved by the oxygen radical absorbance capacity (ORAC) and total radical trapping antioxidant parameter (TRAP). The second generally includes DPPH and ABTS radicals scavenging assays, Trolox equivalent antioxidant capacity (TEAC) (based on the use ABTS), and ferric reducing antioxidant power (FRAP), among others [100]. However, the classification of methods is difficult due to the simultaneous occurrence of both mechanisms in widely utilized antioxidant methods such as DPPH and TEAC [19]. Although the use of these *in vitro* assays has generated controversy as the results cannot be extrapolated to *in vivo* (human) effects [19, 100], these methods are still of great importance as a first screening tool. Among them, ORAC has emerged as an assay method of choice in measuring the capacity to scavenge peroxyl radicals, which are relevant to *in vivo* conditions [33, 100]. Some recent evidences suggest that ORAC is not only a measure of the capacity of antioxidants to trap peroxyl radicals (ROO•), but also to scavenge alkoxyl radicals (RO•) [101]. Antioxidant peptides have other uses apart from what takes place in the human body since protein hydrolyzates can be employed as natural antioxidants to slow oxidations and prevent rancidity in food [102].

Concerning studies on peptides from underused plant resources, antioxidant peptides have been produced from broken rice byproducts [13, 34], brewer's spent grain [12], olive and plum seed wastes [20, 21] and potato byproducts [16] using peptidases of plant, animal and microbial origin. However, better results have been obtained using microbial enzymes or complete cells suspensions (Tables **2** and **4**). This is in line with results on other protein sources [33].

Moreover, the use of simulated gastrointestinal digestion on hydrolyzates of brewer's spent grain obtained with Alcalase and Flavourzyme did not change the ORAC activity, while it changed the DPP-IV and ACE inhibitory activities. Although different results can be obtained depending on the enzyme and biomass applied, using this approach, the latter authors identified two antioxidant peptides, IPY and LPY [12] (Table **5**). In this sense, aromatic amino acids have been described by different authors as the main driver of antioxidant activity of peptides [33, 73]. Recently, Dei Piu' *et al.* [13] have characterized Tyr, Trp, sulfur-containing peptides in antioxidant fractions from rice byproduct and most of them from glutelins (Table **5**). Besides aromatic amino acids, the latter amino acids also contribute to the antioxidant activity of peptides. Moreover, His-containing peptides were also present in some fractions. This amino acid has shown antioxidant properties due to its imidazole ring, which acts as a donor and acceptor of protons as well as a metal ion-chelating agent [13, 103]. In this regard, Esteve *et al.* [20] screened several microbial enzymes (Alcalase 2.4 L FG, Neutrase 0.8 L, Flavourzyme 1000 L, and PTN 6.0 Saltfree*N and thermolysin) on the hydrolysis of olive seeds. The best results were obtained by Alcalase, and several peptides containing Tyr, His and Phe were characterized in the 3 kDa fraction (Table **5**).

Antihypertensive Activity

Hypertension is an important public health challenge worldwide. Firstly, untreated hypertension can lead to cardiovascular and renal diseases, such as stroke, coronary heart disease, and kidney dysfunction [104]. Secondly, hypertension forms part of the metabolic syndrome cluster [105]. Therefore, there is a growing interest in the development of new functional foods and nutraceuticals with the antihypertensive activity that could be useful for the treatment of borderline to mild high blood pressure that does not warrant the prescription of antihypertensive drugs [104]. Current research has focused on the production of novel antihypertensive peptides, mainly from food proteins, but agricultural and agroindustrial byproducts are an alternative source that do not compete with human or even animal nutrition (Tables **2** and **4**).

As a first screening assay, the inhibition of the angiotensin I-converting enzyme (ACE), which could play a pivotal role in blood pressure regulation, can be helpful. ACE catalyzes the production of a vasoconstrictor, angiotensin II, and inactivates the vasodilator, bradykinin. This enzyme is mainly found as a membrane-bound ectoenzyme in vascular endothelial cells [88]. Therefore, ACE-inhibitors are able to reduce the hydrolysis rate of ACE substrates in *in vitro* assays, such as hippuryl-histidyl-leucine, including the spectrophotometric method of Cushman and Cheung, *o*-aminobenzoylglycyl-*p*-nitrophenylalany-lproline and furanacryloyl [106]. However, it is difficult to establish a direct relationship between ACE-inhibitory activity in *vitro* and antihypertensive activity *in vivo* [106].

In this context, several studies have applied hydrolysis to generate ACE inhibitors from underused plant food byproducts. Using Alcalase and thermolysin, hydrolyzates with ACE-inhibitory properties have been obtained from wheat germ protein [42], canola proteins [50], brewer's spent grain [25], cauliflower leaves

[14], as well as grape [53], plum [21], peach [26] and olive seeds [20] Table **2** and **4**. The results provided by Connolly *et al.* [29] suggested that hydrolyzates generated with *Bacillus licheniformis* derived proteinases, including Alcalase, inhibited ACE to a greater extent than hydrolyzates generated with enzymes from other sources, including digestive ones as Corolase PP enzyme preparations. Generally, the most effective food ACE inhibitory peptides are reported to contain Tyr, Phe, Trp and/or Pro at the C terminal [18, 29] (see Table **5**). For example, Alcalase, which mainly contains subtilisin endoproteinase with minor glutamyl endopeptidase, can generate ACEi peptides containing Pro [29]. Moreover, Cermeño *et al.* [12] have evidenced that, although Ile and Pro containing peptides, derived from brewer's spent grain, had higher *in vitro* bioactivities when compared to peptides having Leu and Pro, the length is also important. Furthermore, autolysis of potato byproducts provided a way to obtain ACEi activity, but it was less effective than enzymatic hydrolysis with *Bacillus* peptidases [16].

Interestingly, the IC_{50} values of some of these hydrolyzates are in the range of those found after casein hydrolysis, a common source of bioactive peptides with also antihypertensive activity *in vivo* [107] and humans [108]. The best values were for canola and peach seed hydrolyzates, with IC_{50} values of 0.015 and 0.016 mg/mL, respectively.

Recently, Cermeño *et al.* [12] also showed that a protein hydrolyzate from brewer's spent grain, produced using Alcalase and Flavourzyme, had blood pressure-reducing properties in animal models. It promoted reduction values of 37.8 for SBP, 24.14 for DBP, and 31.93 for MAP after 6 h post administration. Some active sequences were identified after simulated gastrointestinal digestion, including ACE inhibitors like IPLQP and LPLQP (Table **5**).

Diabetes Targets

The prevalence of type 2 diabetes is increasing in the last number of years. Moreover, metabolic syndrome has been associated with an increased risk of cardiovascular diseases and type 2 diabetes [105]. In this context, the inhibition of dipeptidyl peptidase IV (DPP-IV) could be a strategy for the management of type II diabetes. This enzyme degrades incretins, which enhance insulin secretion from pancreatic beta cells in the presence of glucose *in vivo* [109]. In this way, Connolly and coworkers have evaluated a SeEH strategy to produce DPP-IV inhibitors from brewer's spent grain protein enriched isolate. Among the enzymes tested, the porcine pancreatic proteinase preparation Corolase PP produced hydrolyzates with the highest DPP-IV inhibitory potential, but resulted in relatively low ACE inhibition [29].

Another target could be through slowing the action of gastrointestinal carbohydrolases in order to prolong the digestion time of carbohydrates and decrease the rate of glucose absorption [105]. In this way, Connolly *et al.* [29] also evaluated the inhibition of α-glucosidase and α-amylase. While tryptic digests of brewer's spent grain proteins resulted in the greatest α-glucosidase inhibitory activity, no significant change in α-amylase inhibition was observed with hydrolyzates obtained by Corolase PP and others enzyme preparations derived from papaya with papain activity (Corolase L10, and Promod 144MG), *Aspergillus oryzae* (Flavourzyme 500) and *Bacillus licheniformis* (Alcalase 2.4 L, Protex6L, Protamex, Prolyve 1000, *etc.*).

Moreover, cauliflower leaves protein hydrolysed by Alcalase promoted the glucose consumption and enhanced the glycogen content in human liver cancer cell line HepG2, which suggests an ability to regulate glucose metabolism [14] (Table **2**).

Antimicrobial Activity

There is a growing interest in the food industry and among consumers in reducing the use of synthetic preservatives in food preservation and opting instead for natural preservatives, including antioxidant and antimicrobial agents. An example is nisin (E234), a polycyclic peptide of microbial origin used as food preservative [110]. In this regard, obtaining plant antimicrobial peptides could be a way to obtain natural antimicrobial agents. Concerning agroindustrial byproducts, the hydrolysis of proteins from palm kernel cake, which were extracted using a strong alkaline solution, by Alcalase were subjected to further purification to recover the active antimicrobial fraction [23, 111] (Table **2**). It was not a pure peptide but instead a peptide-containing compound high in lauric acid derivative [111].

Other Bioactivities

Other bioactivities have been reported for plant protein hydrolyzates derived from agricultural and agroindustrial byproducts (Tables **2** and **4**). In this sense, protein hydrolyzates from walnut, canola and rice byproducts have exhibited: growth inhibitory activity on cancer cell lines, bile acid-binding capacity to reduce low-density lipoprotein cholesterol, anti-thrombotic activity, and protection against UV radiation [15, 112, 113]. In particular, Ma *et al.* [15] isolated the anticancer peptide CTLEW after the hydrolysis of walnut residue with papain, an active endolytic cysteine protease that exhibits a preference in an amino acid bearing a large hydrophobic side chain. For that, these authors applied sequential ultra-filtration, gel filtration chromatography and RP-HPLC. This peptide was characterized by an amphiphilic structure that might be conducive to invade cancer cell membranes.

Besides the aforementioned bioactivities reported for brewer's spent grain hydrolyzates, a recent study by Cian *et al.* [24] revealed that this byproduct, after being treated with carbohydrolases and a combination of peptidases, showed antithrombotic activity by inhibiting the intrinsic and common pathways of blood coagulation. Moreover, after simulated gastrointestinal digestion, brewer's spent grain peptides were able to delay clotting time, showed higher α-glucosidase and α-amylase activity, but lower ACEi activity. The latter could be related to a lower Pro content. Moreover, immunomodulatory activity was found when alkaline extracted brewer's spent grain protein-rich fraction was hydrolyzed with Prolyve/Protease P and Corolase PP/Flavourzyme; *i.e.* a 58.30% and 48.02% reduction in the release of interleukin-6 was found, respectively, when compared to untreated LPS stimulated control cells (100%) [35].

Recently, Ferri *et al.* [62] have reported rice broken byproduct-derived peptides with anti-tyrosinase, antioxidant and anti-hypertensive activities and poor cytotoxicity and irritation capacities. The highest levels of bioactivity were detected in Protamex-derived samples, followed by samples treated with Alcalase, and compared to Neutrase, while Flavourzyme and papain did not induce peptide release as efficiently as the other three enzymes. Moreover, Protamex hydrolyzate also showed anti-inflammatory activity, which was evaluated as the capability of samples to inhibit TNFα-induced NF-κB activity in a cell-based assay.

SAFETY ISSUES

As commented before, protein hydrolyzates from plant byproducts can be applied for animal feeding as well as to formulate food additives and functional ingredients for foods, nutraceuticals and cosmetics. Moreover, protein hydrolysis can be employed to improve the techno-functional properties of the parent proteins, acquiring new properties: emulsifying, foaming and dispersibility properties, water binding capacity, being bulk agents for new materials, *etc.*

However, for human nutrition, there are some constraints. Firstly, in the EU the industrial application of protein hydrolyzates from agricultural and agroindustrial byproducts may be hindered by legislative issues, especially in the case of proteins from those plants not commonly used as food. In this sense, the respective Novel Food Regulation (Regulation (EU) 2015/2283) [114] classifies food obtained in a production process not used for food production before 15 May 1997, as "novel food" and the regulatory status for each single case must be gathered [32]. In this sense, EU countries may restrict the marketing of a product through specific legislation and businesses should address their national authorities. Even if foods and/or food ingredients were used in supplements for food, new food uses require authorization. In addition, GRAS status is required for food enzymes; *e.g.* those just accepted can be consulted in the websites of the European Union (https://www.efsa.europa.eu/en/topics/topic/food-enzymes; accessed January 17, 2020) and U.S. Food and Drug Administration (https://www.fda.gov/food/generally-recognized-safe-gras/enz-me-preparations-used-food-partial-list; accessed January 17, 2020).

Nonetheless, the future is in waste as commented in the introduction. According to FAO, food waste is estimated to be 1.3 billion tons annually from field to fork, which is equivalent to around one-third of global food production. This number is expected to increase in parallel with further economic and global population increases [115]. Currently, Europe's Bioeconomy Strategy [116] addresses the production of renewable biological resources and their conversion into bio-products and bioenergy, looking for a "zero waste". This could be an opportunity for protein hydrolyzates from these bioresources, especially in a biorefinery framework, as noted in the introduction section. Moreover, the potential use of protein hydrolyzates as food ingredients may satisfy new consumer preferences for natural additives, as commented before.

In any case, the protein product has to be safe. This means, for example, low toxicity. From a nutritional point of view, low denaturation of essential amino acids is desirable. Thus, relatively mild protein production is preferable for food applications [41]. In this sense, there are several mechanisms that can have an effect on amino acids due to thermal processing: racemization, amino acid decomposition and formation of new molecules, glycation (nonenzymatic browning or Maillard reaction), and cross-linking, *e.g.* formation of lysinoalanine (LAL), *etc* [107]. LAL can be formed during alkali treatment of protein and it induces cytomegaly or karyomegaly in the kidney of certain animals like rats and mice [117]. The formation and the mechanisms of LAL formation/decomposition are not generally studied [6] and thus, it requires further attention. For example, Hou *et al.* [93] revealed that an increase of LAL content was observed up to 0.09 M NaOH treatment, and then it decreased. Using high temperature and pressure, *e.g.* using subcritical conditions, the degradation of amino acids to low molecular weight carboxylic acids, such as formic acid, acetic acid, and propionic acids, may also occur [118]. Interestingly, SiSH methods can improve protein recovery (even, higher than 80%), and moderate the thermal-alkaline requirements [6]. Due to the reduction in protein size through proteolysis, the proteins can be more easily extracted and this allows a lowering of processing pH, thus avoiding severe conditions that denature protein [41] or other plant constituents.

Sometimes, protein flours and concentrates can be accompanied by the presence of undesirable compounds, such as anti-nutritional and toxic compounds [37], or even toxic proteins like those found in castor bean plant (*Ricinus communis* L.)

[119]. Accordingly, the content of these compounds should be limited and monitored if the protein product is destined for nutrition. In other cases, the presence of other phytochemicals could be desirable to obtain multifunctional ingredients or to modulate antioxidant activity [94]. Accordingly, previous commercialization toxicology studies are necessary to assess the safety or, alternatively, the possible adverse effects that might be exerted by the peptides, their metabolites, or other compounds present in the peptide product [120].

CONCLUSION

Most studies on agricultural and agroindustrial byproducts have applied SeEH methods to recover, firstly, protein products and then to produce bioactive hydrolyzates. Nonetheless, direct SiEH methods could be applied alternatively with environmental and economic advantages. This means that SiSH methods require low concentrations of alkali and significantly reduced energy inputs and efforts since it does not require purification/drying steps. Some SeEH methods, consisting of a previous pretreatment with carbohydrolases before proteolysis, are in the interface, when no further purification steps are applied. In this sense, some studies suggest that this pretreatment can enhance the technical yield. Finally, as various pharmacological activities of protein hydrolyzates have been reported here, they can be therefore therapeutically beneficial. Other application could be as food preservatives. Nevertheless, as far as we know, these studies have been primarily tested *in vitro*, with some exceptions. This means that the assessment of the reported bioactivity *in vivo* should be performed for a real application as functional ingredients together with clinical evidence and safety studies.

PATIENT CONSENT

It is not required.

CONSENT FOR PUBLICATION

Not applicable.

CONFLICT OF INTEREST

The authors confirm that this chapter content has no conflict of interest.

ACKNOWLEDGEMENTS

The authors are grateful for the postdoctoral grants funded by "Acción 6 del Plan de Apoyo a la Investigación de la Universidad de Jaén (UJA), 2017–2019" and FEDER UJA projects 1260905 and 1262115 funded by "Programa Operativo FEDER 2014-2020" and "Consejería de Economía y Conocimiento de la Junta de

Andalucía". Also, we thank the grant "Beca-comisión para estadía técnica del Tecnológico Nacional de México/Instituto Tecnológico de Zitácuaro y Universidad de Jaén".

REFERENCES

[1]   Unite Nations. *Goal 2: End hunger, achieve food security and improved nutrition and promote sustainable agriculture,* https://unstats.un.org/sdgs/report/2016/Goal-02/

[2]   FAO. *IFAD, UNICEF, WFP and WHO. The State of Food Security and Nutrition in the World 2018. Building climate resilience for food security and nutrition*; FAO: Rome, **2018**, pp. 1-182.

[3]   Gortmaker, S.L.; Swinburn, B.A.; Levy, D.; Carter, R.; Mabry, P.L.; Finegood, D.T.; Huang, T.; Marsh, T.; Moodie, M.L. Changing the future of obesity: science, policy, and action. *Lancet,* **2011**, *378*(9793), 838-847.
      [http://dx.doi.org/10.1016/S0140-6736(11)60815-5] [PMID: 21872752]

[4]   Henchion, M.; Hayes, M.; Mullen, A.M.; Fenelon, M.; Tiwari, B. Future Protein supply and demand: Strategies and factors influencing a sustainable equilibrium. *Foods,* **2017**, *6*(7), 53.
      [http://dx.doi.org/10.3390/foods6070053] [PMID: 28726744]

[5]   Communication from the Commission to the European Parliament. *The Council, the European Economic and Social Committee and the Committee of the Regions. COM,* **2011**, 615.

[6]   Contreras, M.D.M.; Lama-Muñoz, A.; Manuel Gutiérrez-Pérez, J.; Espínola, F.; Moya, M.; Castro, E. Protein extraction from agri-food residues for integration in biorefinery: Potential techniques and current status. *Bioresour. Technol.,* **2019**, *280*(280), 459-477.
      [http://dx.doi.org/10.1016/j.biortech.2019.02.040] [PMID: 30777702]

[7]   Contreras, M.D.M.; Lama-Muñoz, A.; Gutiérrez-Pérez, J.M.; Espínola, F.; Moya, M.; Romero, I.; Castro, E. Integrated process for sequential extraction of bioactive phenolic compounds and proteins from mill and field olive leaves and effects on the lignocellulosic profile. *Foods,* **2019**, *8*(11)E531
      [http://dx.doi.org/10.3390/foods8110531] [PMID: 31671747]

[8]   Contreras, M.D.M.; Lama-Muñoz, A.; Espínola, F.; Moya, M.; Romero, I.; Castro, E. Valorization of olive mill leaves through ultrasound-assisted extraction. *Food Chem.,* **2020**, *314*126218
      [http://dx.doi.org/10.1016/j.foodchem.2020.126218] [PMID: 31982857]

[9]   Kudic, M.; Shkolnykova, M. From biotech to bioeconomy. *Bremen papers on economics & innovation*; Günther, J., Ed.; University of Bremen: Bremen, **2019**, pp. 1-41.

[10] Daliri, E.B-M.; Oh, D.H.; Lee, B.H. Bioactive Peptides. *Foods,* **2017**, *6*(5), 32.
      [http://dx.doi.org/10.3390/foods6050032] [PMID: 28445415]

[11] Sharifi-Rad, M.; Ozcelik, B.; Altın, G.; Daşkaya-Dikmen, C.; Martorell, M.; Ramírez-Alarcón, K.; Alarcón-Zapata, P.; Morais-Braga, M.F.B.; Carneiro, J.N.P.; Alves Borges Leal, A.L. *Salvia* spp. plants-from farm to food applications and phytopharmacotherapy. *Trends Food Sci. Technol.,* **2018**, *80*, 242-263.
      [http://dx.doi.org/10.1016/j.tifs.2018.08.008]

[12] Cermeño, M.; Connolly, A.; O'Keeffe, M.B.; Flynn, C.; Alashi, A.M.; Aluko, R.E.; FitzGerald, R.J. Identification of bioactive peptides from brewers' spent grain and contribution of Leu/Ile to bioactive potency. *J. Funct. Foods,* **2019**, *60*(July)
      [http://dx.doi.org/10.1016/j.jff.2019.103455]

[13] Dei Piu', L.; Tassoni, A.; Serrazanetti, D.I.; Ferri, M.; Babini, E.; Tagliazucchi, D.; Gianotti, A. Exploitation of starch industry liquid by-product to produce bioactive peptides from rice hydrolyzed proteins. *Food Chem.,* **2014**, *155*, 199-206.
      [http://dx.doi.org/10.1016/j.foodchem.2014.01.055] [PMID: 24594175]

[14] Xu, Y.; Li, Y.; Bao, T.; Zheng, X.; Chen, W.; Wang, J. A recyclable protein resource derived from

cauliflower by-products: Potential biological activities of protein hydrolysates. *Food Chem.,* **2017,** *221,* 114-122.
[http://dx.doi.org/10.1016/j.foodchem.2016.10.053] [PMID: 27979071]

[15]   Ma, S.; Huang, D.; Zhai, M.; Yang, L.; Peng, S.; Chen, C.; Feng, X.; Weng, Q.; Zhang, B.; Xu, M. Isolation of a novel bio-peptide from walnut residual protein inducing apoptosis and autophagy on cancer cells. *BMC Complement. Altern. Med.,* **2015,** *15*(1), 413.
[http://dx.doi.org/10.1186/s12906-015-0940-9] [PMID: 26593407]

[16]   Pihlanto, A.; Akkanen, S.; Korhonen, H.J. ACE-inhibitory and antioxidant properties of potato (*Solanum tuberosum*). *Food Chem.,* **2008,** *109*(1), 104-112.
[http://dx.doi.org/10.1016/j.foodchem.2007.12.023] [PMID: 26054270]

[17]   Sari, Y.W.; Mulder, W.J.; Sanders, J.P.M.; Bruins, M.E. Towards plant protein refinery: Review on protein extraction using alkali and potential enzymatic assistance. *Biotechnol. J.,* **2015,** *10*(8), 1138-1157.
[http://dx.doi.org/10.1002/biot.201400569] [PMID: 26132986]

[18]   Hernández-Ledesma, B.; Contreras, Md.M.; Recio, I. Antihypertensive peptides: production, bioavailability and incorporation into foods. *Adv. Colloid Interface Sci.,* **2011,** *165*(1), 23-35.
[http://dx.doi.org/10.1016/j.cis.2010.11.001] [PMID: 21185549]

[19]   Lorenzo, J.M.; Munekata, P.E.S.; Gómez, B.; Barba, F.J.; Mora, L.; Pérez-Santaescolástica, C.; Toldrá, F. Bioactive peptides as natural antioxidants in food products – A review. *Trends Food Sci. Technol.,* **2018,** *79*(July), 136-147.
[http://dx.doi.org/10.1016/j.tifs.2018.07.003]

[20]   Esteve, C.; Marina, M.L.; García, M.C. Novel strategy for the revalorization of olive (*Olea europaea*) residues based on the extraction of bioactive peptides. *Food Chem.,* **2015,** *167,* 272-280.
[http://dx.doi.org/10.1016/j.foodchem.2014.06.090] [PMID: 25148989]

[21]   González-García, E.; Marina, M.L.; García, M.C. Plum (*Prunus Domestica* L.) by-product as a new and cheap source of bioactive peptides: Extraction method and peptides characterization. *J. Funct. Foods,* **2014,** *11*(C), 428-437.
[http://dx.doi.org/10.1016/j.jff.2014.10.020]

[22]   Hamada, J.S. Characterization and functional properties of rice bran proteins modified by commercial exoproteases and endoproteases. *J. Food Sci.,* **2000,** *65*(2), 305-310.
[http://dx.doi.org/10.1111/j.1365-2621.2000.tb15998.x]

[23]   Tan, Y.N.; Ayob, M.K.; Osman, M.A.; Matthews, K.R. Antibacterial activity of different degree of hydrolysis of palm kernel expeller peptides against spore-forming and non-spore-forming bacteria. *Lett. Appl. Microbiol.,* **2011,** *53*(5), 509-517.
[http://dx.doi.org/10.1111/j.1472-765X.2011.03137.x] [PMID: 21848644]

[24]   Cian, R.E.; Garzón, A.G.; Martínez-Augustin, O.; Botto, C.C.; Drago, S.R. Antithrombotic activity of brewers' spent grain peptides and their effects on blood coagulation pathways. *Plant Foods Hum. Nutr.,* **2018,** *73*(3), 241-246.
[http://dx.doi.org/10.1007/s11130-018-0682-1] [PMID: 29992417]

[25]   Connolly, A.; O'Keeffe, M.B.; Piggott, C.O.; Nongonierma, A.B.; FitzGerald, R.J. Generation and identification of angiotensin converting enzyme (ACE) inhibitory peptides from a brewers' spent grain protein isolate. *Food Chem.,* **2015,** *176,* 64-71.
[http://dx.doi.org/10.1016/j.foodchem.2014.12.027] [PMID: 25624207]

[26]   Vásquez-Villanueva, R.; Marina, M.L.; García, M.C. Revalorization of a peach (*Prunus persica* (L.) Batsch) byproduct: Extraction and characterization of ACE-Inhibitory peptides from peach stones. *J. Funct. Foods,* **2015,** *18,* 137-146.
[http://dx.doi.org/10.1016/j.jff.2015.06.056]

[27]   Contreras, M.M.; Sánchez, D.; Sevilla, M.A.; Recio, I.; Amigo, L. Resistance of casein-derived bioactive peptides to simulated gastrointestinal digestion. *Int. Dairy J.,* **2013,** *32,* 71-78.

[http://dx.doi.org/10.1016/j.idairyj.2013.05.008]

[28] Merz, M.; Eisele, T.; Berends, P.; Appel, D.; Rabe, S.; Blank, I.; Stressler, T.; Fischer, L. Flavourzyme, an enzyme preparation with industrial relevance: Automated nine-step purification and partial characterization of eight enzymes. *J. Agric. Food Chem.,* **2015**, *63*(23), 5682-5693.
[http://dx.doi.org/10.1021/acs.jafc.5b01665] [PMID: 25996918]

[29] Connolly, A.; Piggott, C.O.; FitzGerald, R.J. *In vitro* α-Glucosidase, angiotensin converting enzyme and dipeptidyl peptidase-iv inhibitory properties of brewers' spent grain protein hydrolysates. *Food Res. Int.,* **2014**, *56*, 100-107.
[http://dx.doi.org/10.1016/j.foodres.2013.12.021]

[30] Geng, J.Z.; Shao, J.; Yang, J.H.; Pang, B.; Cao, C.X.; Fan, L.Y. Reassemblable quasi-chip free-flow electrophoresis with simple heating dispersion for rapid micropreparation of trypsin in crude porcine pancreatin. *Electrophoresis,* **2011**, *32*(22), 3248-3256.
[http://dx.doi.org/10.1002/elps.201100358] [PMID: 22102499]

[31] Martínez-Maqueda, D.; Hernández-Ledesma, B.; Amigo, L.; Miralles, B.; Gómez-Ruiz, J.A. Extraction/fractionation tecniques for proteins and peptides and protein digestion.*Proteomics in Foods: Principles and applications*; Todrá, F.; Nollet, L.M.L., Eds.; Springer US: New York, **2013**, pp. 21-50.
[http://dx.doi.org/10.1007/978-1-4614-5626-1_2]

[32] Pojić, M.; Mišan, A.; Tiwari, B. Eco-innovative technologies for extraction of proteins for human consumption from renewable protein sources of plant origin. *Trends Food Sci. Technol.,* **2018**, *75*(March), 93-104.
[http://dx.doi.org/10.1016/j.tifs.2018.03.010]

[33] Contreras, M.M.; Hernández-Ledesma, B.; Amigo, L.; Martín-Álvarez, P.J.; Recio, I. Production of antioxidant hydrolyzates from a whey protein concentrate with thermolysin: Optimization by response surface methodology. *Lebensm. Wiss. Technol.,* **2011**, *44*(1), 9-15.
[http://dx.doi.org/10.1016/j.lwt.2010.06.017]

[34] Rani, S.; Pooja, K.; Kumar, G. Exploration of rice protein hydrolysates and peptides with special reference to antioxidant potential : Computational derived approaches for bio-activity determination. *Trends Food Sci. Technol.,* **2018**, *80*, 61-70.
[http://dx.doi.org/10.1016/j.tifs.2018.07.013]

[35] Connolly, A.; Cermeño, M.; Crowley, D.; O'Callaghan, Y.; O'Brien, N.M.; FitzGerald, R.J. Characterisation of the *in vitro* bioactive properties of alkaline and enzyme extracted brewers' spent grain protein hydrolysates. *Food Res. Int.,* **2019**, *121*(121), 524-532.
[http://dx.doi.org/10.1016/j.foodres.2018.12.008] [PMID: 31108777]

[36] Qin, F.; Johansen, A.Z.; Mussatto, S.I. Evaluation of different pretreatment strategies for protein extraction from brewer's spent grains. *Ind. Crops Prod.,* **2018**, *125*(September), 443-453.
[http://dx.doi.org/10.1016/j.indcrop.2018.09.017]

[37] Rodrigues, I.M.; Coelho, J.F.J.; Carvalho, M.G.V.S. Isolation and valorisation of vegetable proteins from oilseed plants: methods, limitations and potential. *J. Food Eng.,* **2012**, *109*(3), 337-346.
[http://dx.doi.org/10.1016/j.jfoodeng.2011.10.027]

[38] Montealegre, C.; Esteve, C.; García, M.C.; García-Ruiz, C.; Marina, M.L. Proteins in olive fruit and oil. *Crit. Rev. Food Sci. Nutr.,* **2014**, *54*(5), 611-624.
[http://dx.doi.org/10.1080/10408398.2011.598639] [PMID: 24261535]

[39] Connolly, A.; Piggott, C.O.; Fitzgerald, R.J. Characterisation of protein-rich isolates and antioxidative phenolic extracts from pale and black brewers' spent grain. *Int. J. Food Sci. Technol.,* **2013**, *48*(8), 1670-1681.
[http://dx.doi.org/10.1111/ijfs.12137]

[40] Hamada, J.S. Characterization of protein fractions of rice bran to devise effective methods of protein solubilization. *Cereal Chem.,* **1997**, *74*(5), 662-668.

[http://dx.doi.org/10.1094/CCHEM.1997.74.5.662]

[41]   Sari, Y.W.; Syafitri, U.; Sanders, J.P.M.; Bruins, M.E. How biomass composition determines protein extractability. *Ind. Crops Prod.,* **2015**, *70*, 125-133.
[http://dx.doi.org/10.1016/j.indcrop.2015.03.020]

[42]   Jia, J.; Ma, H.; Zhao, W.; Wang, Z.; Tian, W.; Luo, L.; He, R. The use of ultrasound for enzymatic preparation of ACE-inhibitory peptides from wheat germ protein. *Food Chem.,* **2010**, *119*(1), 336-342.
[http://dx.doi.org/10.1016/j.foodchem.2009.06.036]

[43]   Tang, S.; Hettiarachchy, N.S.; Eswaranandam, S.; Crandall, P. Protein extraction from heat-stabilized defatted rice bran: II. The role of Amylase, Celluclast, and Viscozyme. *J. Food Sci.,* **2003**, *68*(2), 471-475.
[http://dx.doi.org/10.1111/j.1365-2621.2003.tb05696.x]

[44]   Rommi, K.; Niemi, P.; Kemppainen, K.; Kruus, K. Impact of thermochemical pre-treatment and carbohydrate and protein hydrolyzing enzyme treatment on fractionation of protein and lignin from brewer's spent grain. *J. Cereal Sci.,* **2018**, *79*, 168-173.
[http://dx.doi.org/10.1016/j.jcs.2017.10.005]

[45]   Treimo, J.; Aspmo, S.I.; Eijsink, V.G.H.; Horn, S.J. Enzymatic solubilization of proteins in brewer's spent grain. *J. Agric. Food Chem.,* **2008**, *56*(13), 5359-5365.
[http://dx.doi.org/10.1021/jf073317s] [PMID: 18553975]

[46]   Treimo, J.; Westereng, B.; Horn, S.J.; Forssell, P.; Robertson, J.A.; Faulds, C.B.; Waldron, K.W.; Buchert, J.; Eijsink, V.G.H. Enzymatic solubilization of brewers' spent grain by combined action of carbohydrases and peptidases. *J. Agric. Food Chem.,* **2009**, *57*(8), 3316-3324.
[http://dx.doi.org/10.1021/jf803310f] [PMID: 19284754]

[47]   Zhang, S.B.; Wang, Z.; Xu, S.Y. Optimization of the aqueous enzymatic extraction of rapeseed oil and protein hydrolysates. *JAOCS. J. Am. Oil Chem. Soc.,* **2007**, *84*(1), 97-105.
[http://dx.doi.org/10.1007/s11746-006-1004-6]

[48]   Tang, S.; Hettiarachchy, N.S.; Horax, R.; Eswaranandam, S. Physicochemical properties and functionality of rice bran protein hydrolyzate prepared from heat-stabilized defatted rice bran with the aid of enzymes. *J. Food Sci.,* **2003**, *68*(1), 152-157.
[http://dx.doi.org/10.1111/j.1365-2621.2003.tb14132.x]

[49]   Wang, H.; Wang, J.; Lv, Z.; Liu, Y.; Lu, F. Preparing oligopeptides from broken rice protein by ultrafiltration-coupled enzymatic hydrolysis. *Eur. Food Res. Technol.,* **2013**, *236*(3), 419-424.
[http://dx.doi.org/10.1007/s00217-012-1904-7]

[50]   Wu, J.; Muir, A.D. Comparative structural, emulsifying, and biological properties of 2 major canola proteins, cruciferin and napin. *J. Food Sci.,* **2008**, *73*(3), C210-C216.
[http://dx.doi.org/10.1111/j.1750-3841.2008.00675.x] [PMID: 18387101]

[51]   Esteve, C.; Del Río, C.; Marina, M.L.; García, M.C. First ultraperformance liquid chromatography based strategy for profiling intact proteins in complex matrices: application to the evaluation of the performance of olive (*Olea europaea* L.) stone proteins for cultivar fingerprinting. *J. Agric. Food Chem.,* **2010**, *58*(14), 8176-8182.
[http://dx.doi.org/10.1021/jf101305t] [PMID: 20575522]

[52]   Li, M.; Wen, X.; Peng, Y.; Wang, Y.; Wang, K.; Ni, Y. Functional properties of protein isolates from bell pepper (*Capsicum annuum* L. Var. *Annuum*) seeds. *Lebensm. Wiss. Technol.,* **2017**, *2018*(97), 802-810.
[http://dx.doi.org/10.1016/j.lwt.2018.07.069]

[53]   Ding, Q.; Zhang, T.; Niu, S.; Cao, F.; Wu-Chen, R.A.; Luo, L.; Ma, H. Impact of ultrasound pretreatment on hydrolysate and digestion products of grape seed protein. *Ultrason. Sonochem.,* **2018**, *42*(42), 704-713.
[http://dx.doi.org/10.1016/j.ultsonch.2017.11.027] [PMID: 29429721]

[54]   Fetzer, A.; Herfellner, T.; Stäbler, A.; Menner, M.; Eisner, P. Influence of process conditions during aqueous protein extraction upon yield from pre-pressed and cold-pressed rapeseed press cake. *Ind. Crops Prod.,* **2017**, *2018*(112), 236-246.
       [http://dx.doi.org/10.1016/j.indcrop.2017.12.011]

[55]   Ye, X.; Li, L. Microwave-assisted protein solubilization for mass spectrometry-based shotgun proteome analysis. *Anal. Chem.,* **2012**, *84*(14), 6181-6191.
       [http://dx.doi.org/10.1021/ac301169q] [PMID: 22708679]

[56]   Yan-rong, Z.; Yi, Y.; Chun-chun, S.; College, W. D. W. D. Optimization of preparation process of high f-value corn peptide by microwave-assisted enzymatic hydrolysis. *Food Sci.,* **2013**, *03*

[57]   Ketnawa, S.; Liceaga, A.M. Effect of microwave treatments on antioxidant activity and antigenicity of fish frame protein hydrolysates. *Food Bioprocess Technol.,* **2017**, *10*(3), 582-591.
       [http://dx.doi.org/10.1007/s11947-016-1841-8]

[58]   Cookman, D.J.; Glatz, C.E. Extraction of protein from distiller's grain. *Bioresour. Technol.,* **2009**, *100*(6), 2012-2017.
       [http://dx.doi.org/10.1016/j.biortech.2008.09.059] [PMID: 19028095]

[59]   De Moura, J.M.L.N.; Campbell, K.; Mahfuz, A.; Jung, S.; Glatz, C.E.; Johnson, L. Enzyme-assisted aqueous extraction of oil and protein from soybeans and cream de-emulsification. *JAOCS. J. Am. Oil Chem. Soc.,* **2008**, *85*(10), 985-995.
       [http://dx.doi.org/10.1007/s11746-008-1282-2]

[60]   Lamsal, B.P.; Murphy, P.A.; Johnson, L.A. Flaking and extrusion as mechanical treatments for enzyme-assisted aqueous extraction of oil from soybeans. *JAOCS. J. Am. Oil Chem. Soc.,* **2006**, *83*(11), 973-979.
       [http://dx.doi.org/10.1007/s11746-006-5055-5]

[61]   Wang, B.; Meng, T.; Ma, H.; Zhang, Y.; Li, Y.; Jin, J.; Ye, X. Mechanism study of dual-frequency ultrasound assisted enzymolysis on rapeseed protein by immobilized Alcalase. *Ultrason. Sonochem.,* **2016**, *32*, 307-313.
       [http://dx.doi.org/10.1016/j.ultsonch.2016.03.023] [PMID: 27150775]

[62]   Ferri, M.; Graen-Heedfeld, J.; Bretz, K.; Guillon, F.; Michelini, E.; Calabretta, M.M.; Lamborghini, M.; Gruarin, N.; Roda, A.; Kraft, A.; Tassoni, A. Peptide fractions obtained from rice by-products by means of an environment-friendly process show *in vitro* health-related bioactivities. *PLoS One,* **2017**, *12*(1)e0170954
       [http://dx.doi.org/10.1371/journal.pone.0170954] [PMID: 28125712]

[63]   Latif, S.; Anwar, F. Aqueous enzymatic sesame oil and protein extraction. *Food Chem.,* **2011**, *125*(2), 679-684.
       [http://dx.doi.org/10.1016/j.foodchem.2010.09.064] [PMID: 30634286]

[64]   WHO/FAO. Food energy – methods of analysis and conversion factors. *FAO Food Nutr.,* **2003**, *77*https://doi.org/ISSN

[65]   Rojas-Chamorro, J. A.; Cara, C.; Romero, I.; Ruiz, E.; Romero-García, J. M.; Mussatto, S. I.; Castro, E. Ethanol production from brewers' spent grain pretreated by dilute phosphoric acid. *Energy & Fuels,* **2018**. acs.energyfuels..
       [http://dx.doi.org/https://doi.org/10.1021/acs.energyfuels.8b00343.] [PMID: 8b00343]

[66]   Latif, S.; Pfannstiel, J.; Makkar, H.P.S.; Becker, K. Amino acid composition, antinutrients and allergens in the peanut protein fraction obtained by an aqueous enzymatic process. *Food Chem.,* **2013**, *136*(1), 213-217.
       [http://dx.doi.org/10.1016/j.foodchem.2012.07.120] [PMID: 23017415]

[67]   Görgüç, A.; Özer, P.; Yılmaz, F.M. Microwave-assisted enzymatic extraction of plant protein with antioxidant compounds from the food waste sesame bran: Comparative optimization study and identification of metabolomics using LC/Q-TOF/MS. *J. Food Process. Preserv.,* **2020**, *44*(1), 1-11.

[http://dx.doi.org/10.1111/jfpp.14304]

[68]   Kazan, A.; Celiktas, M.S.; Sargin, S.; Yesil-Celiktas, O. Bio-based fractions by hydrothermal treatment of olive pomace: Process optimization and evaluation. *Energy Convers. Manage.,* **2015,** *103,* 366-373.
[http://dx.doi.org/10.1016/j.enconman.2015.06.084]

[69]   Vergara-Barberán, M.; Lerma-García, M.J.; Herrero-Martínez, J.M.; Simó-Alfonso, E.F. Use of an enzyme-assisted method to improve protein extraction from olive leaves. *Food Chem.,* **2015,** *169,* 28-33.
[http://dx.doi.org/10.1016/j.foodchem.2014.07.116] [PMID: 25236194]

[70]   Celus, I.; Brijs, K.; Delcour, J.A. Enzymatic hydrolysis of brewers' spent grain proteins and technofunctional properties of the resulting hydrolysates. *J. Agric. Food Chem.,* **2007,** *55*(21), 8703-8710.
[http://dx.doi.org/10.1021/jf071793c] [PMID: 17896813]

[71]   Celus, I.; Brijs, K.; Delcour, J.A. Fractionation and characterization of brewers' spent grain protein hydrolysates. *J. Agric. Food Chem.,* **2009,** *57*(12), 5563-5570.
[http://dx.doi.org/10.1021/jf900626j] [PMID: 19456139]

[72]   Thermo Scientific. *Thermo Scientific Pierce Protein Assay Technical Handbook Version 2.,* http://tools.thermofisher.com/content/sfs/brochures/1602063-Protein-Assay-Handbook.pdf

[73]   Contreras, M.M.; Carrón, R.; Montero, M.J.; Ramos, M.; Recio, I. Novel casein-derived peptides with antihypertensive activity. *Int. Dairy J.,* **2009,** *19*(10)
[http://dx.doi.org/10.1016/j.idairyj.2009.05.004]

[74]   Commission (EC) No 152/2009 of 27 January 2009 laying down the methods of sampling and analysis for the official control of feed. *Off. J. Eur. Communities,* **2009,** *L54,* 1-130.

[75]   Talekar, S.; Patti, A.F.; Singh, R.; Vijayraghavan, R.; Arora, A. From waste to wealth: high recovery of nutraceuticals from pomegranate seed waste using a green extraction process. *Ind. Crops Prod.,* **2018,** *112*(January), 790-802.
[http://dx.doi.org/10.1016/j.indcrop.2017.12.023]

[76]   Gerzhova, A.; Mondor, M.; Benali, M.; Aider, M. A comparative study between the electro-activation technique and conventional extraction method on the extractability, composition and physicochemical properties of canola protein concentrates and isolates. *Food Biosci.,* **2016,** *2015*(11), 56-71.
[http://dx.doi.org/10.1016/j.fbio.2015.04.005]

[77]   De Oliveira, A.S.; Campos, J.M.S.; Oliveira, M.R.C.; Brito, A.F.; Valadares Filho, S.C.; Detmann, E.; Valadares, R.F.D.; de Souza, S.M.; Machado, O.L.T. Nutrient digestibility, nitrogen metabolism and hepatic function of sheep fed diets containing solvent or expeller castorseed meal treated with calcium hydroxide. *Anim. Feed Sci. Technol.,* **2010,** *158*(1-2), 15-28.
[http://dx.doi.org/10.1016/j.anifeedsci.2010.02.009]

[78]   Rommi, K.; Ercili-Cura, D.; Hakala, T.K.; Nordlund, E.; Poutanen, K.; Lantto, R. Impact of total solid content and extraction pH on enzyme-aided recovery of protein from defatted rapeseed (*Brassica rapa* L.) press cake and physicochemical properties of the protein fractions. *J. Agric. Food Chem.,* **2015,** *63*(11), 2997-3003.
[http://dx.doi.org/10.1021/acs.jafc.5b01077] [PMID: 25739320]

[79]   Wang, W.; De Dios Alché, J.; Rodríguez-García, M.I. Characterization of olive seed storage proteins. *Acta Physiol. Plant.,* **2007,** *29*(5), 439-444.
[http://dx.doi.org/10.1007/s11738-007-0053-2]

[80]   Yu, D.; Sun, Y.; Wang, W.; O'Keefe, S.F.; Neilson, A.P.; Feng, H.; Wang, Z.; Huang, H. Recovery of protein hydrolysates from brewer's spent grain using enzyme and ultrasonication. *Int. J. Food Sci. Technol.,* **2020,** *55*(1), 357-368.
[http://dx.doi.org/10.1111/ijfs.14314]

[81] Kim, S.T.; Cho, K.S.; Jang, Y.S.; Kang, K.Y. Two-dimensional electrophoretic analysis of rice proteins by polyethylene glycol fractionation for protein arrays. *Electrophoresis,* **2001**, *22*(10), 2103-2109.
[http://dx.doi.org/10.1002/1522-2683(200106)22:10<2103::AID-ELPS2103>3.0.CO;2-W]   [PMID: 11465512]

[82] de Dios Alché, J.; Jiménez-López, J.C.; Wang, W.; Castro-López, A.J.; Rodríguez-García, M.I. Biochemical characterization and cellular localization of 11S type storage proteins in olive (*Olea europaea* L.) seeds. *J. Agric. Food Chem.,* **2006**, *54*(15), 5562-5570.
[http://dx.doi.org/10.1021/jf060203s] [PMID: 16848546]

[83] Vergara-Barberán, M.; Lerma-García, M.J.; Herrero-Martínez, J.M.; Simó-Alfonso, E.F. Use of protein profiles established by CZE to predict the cultivar of olive leaves and pulps. *Electrophoresis,* **2014**, *35*(11), 1652-1659.
[http://dx.doi.org/10.1002/elps.201300530]

[84] Zhu, Z.; Lu, J.J.; Liu, S. Protein separation by capillary gel electrophoresis: a review. *Anal. Chim. Acta,* **2012**, *709*, 21-31.
[http://dx.doi.org/10.1016/j.aca.2011.10.022] [PMID: 22122927]

[85] Vergara-Barberán, M.; Lerma-García, M.J.; Herrero-Martínez, J.M.; Simó-Alfonso, E.F. Classification of olive leaves and pulps according to their cultivar by using protein profiles established by capillary gel electrophoresis. *Anal. Bioanal. Chem.,* **2014**, *406*(6), 1731-1738.
[http://dx.doi.org/10.1007/s00216-013-7585-7] [PMID: 24390412]

[86] Contreras, Mdel.M.; López-Expósito, I.; Hernández-Ledesma, B.; Ramos, M.; Recio, I. Application of mass spectrometry to the characterization and quantification of food-derived bioactive peptides. *J. AOAC Int.,* **2008**, *91*(4), 981-994.
[http://dx.doi.org/10.1093/jaoac/91.4.981] [PMID: 18727560]

[87] Connolly, A.; O'Keeffe, M.B.; Nongonierma, A.B.; Piggott, C.O.; FitzGerald, R.J. Isolation of peptides from a novel brewers spent grain protein isolate with potential to modulate glycaemic response. *Int. J. Food Sci. Technol.,* **2017**, *52*(1), 146-153.
[http://dx.doi.org/10.1111/ijfs.13260]

[88] Yamamoto, N.; Ejiri, M.; Mizuno, S. Biogenic peptides and their potential use. *Curr. Pharm. Des.,* **2003**, *9*(16), 1345-1355.
[http://dx.doi.org/10.2174/1381612033454801] [PMID: 12769742]

[89] Contreras, M.d.M.; Arráez-Román, D.; Fernández-Gutiérrez, A.; Segura-Carretero, A. Nano-liquid chromatography coupled to time-of-flight mass spectrometry for phenolic profiling: a case study in cranberry syrups. *Talanta,* **2015**, *132*, 929-938.
[http://dx.doi.org/10.1016/j.talanta.2014.10.049] [PMID: 25476399]

[90] Contreras, M.d.M; Borrás-Linares, I.; Herranz-López, M.; Micol, V.; Segura-Carretero, A. Further exploring the absorption and enterocyte metabolism of quercetin forms in the Caco-2 model using nano-LC-TOF-MS. *Electrophoresis,* **2016**, *37*(7-8), 998-1006.
[http://dx.doi.org/10.1002/elps.201500375] [PMID: 26542615]

[91] Esteve, C.; D'Amato, A.; Marina, M.L.; García, M.C.; Citterio, A.; Righetti, P.G. Identification of olive (*Olea europaea*) seed and pulp proteins by nLC-MS/MS *via* combinatorial peptide ligand libraries. *J. Proteomics,* **2012**, *75*(8), 2396-2403.
[http://dx.doi.org/10.1016/j.jprot.2012.02.020] [PMID: 22387115]

[92] González-García, E.; Marina, M.L.; García, M.C.; Righetti, P.G.; Fasoli, E. Identification of plum and peach seed proteins by nLC-MS/MS *via* combinatorial peptide ligand libraries. *J. Proteomics,* **2016**, *148*, 105-112.
[http://dx.doi.org/10.1016/j.jprot.2016.07.024] [PMID: 27469892]

[93] Hou, F.; Ding, W.; Qu, W.; Oladejo, A.O.; Xiong, F.; Zhang, W.; He, R.; Ma, H. Alkali solution extraction of rice residue protein isolates: Influence of alkali concentration on protein functional,

structural properties and lysinoalanine formation. *Food Chem.,* **2017**, *218*, 207-215.
[http://dx.doi.org/10.1016/j.foodchem.2016.09.064] [PMID: 27719899]

[94]   Silva, F.G.D.; Hernández-Ledesma, B.; Amigo, L.; Netto, F.M.; Miralles, B. Identification of peptides released from flaxseed *(Linum usitatissimum)* protein by Alcalase® hydrolysis: Antioxidant activity. *Lebensm. Wiss. Technol.,* **2016**, *2017*(76), 140-146.
[http://dx.doi.org/10.1016/j.lwt.2016.10.049]

[95]   Guil-Guerrero, J.L.; Ramos, L.; Moreno, C.; Zúñiga-Paredes, J.C.; Carlosama-Yepez, M.; Ruales, P. Antimicrobial activity of plant-food by-products: A review focusing on the tropics. *Livest. Sci.,* **2016**, *189*, 32-49.
[http://dx.doi.org/10.1016/j.livsci.2016.04.021]

[96]   Pina-Pérez, M.C.; Ferrús Pérez, M.A. Antimicrobial potential of legume extracts against foodborne pathogens: A review. *Trends Food Sci. Technol.,* **2018**, *72*, 114-124.
[http://dx.doi.org/10.1016/j.tifs.2017.12.007]

[97]   Marciniak, A.; Suwal, S.; Naderi, N.; Pouliot, Y.; Doyen, A. Enhancing enzymatic hydrolysis of food proteins and production of bioactive peptides using high hydrostatic pressure technology. *Trends Food Sci. Technol.,* **2018**, *80*, 187-198.
[http://dx.doi.org/10.1016/j.tifs.2018.08.013]

[98]   Guo, L.; Fitzgerald, R.J. Food protein-derived chelating peptides : biofunctional ingredients for dietary mineral bioavailability enhancement. *Trends Food Sci. Technol.,* **2014**, *37*, 92-105.
[http://dx.doi.org/10.1016/j.tifs.2014.02.007]

[99]   Ozuna, C.; León-galván, M.F. Cucurbitaceae seed protein hydrolysates as a potential source of bioactive peptides with functional properties. *Biomed Res. Int.***2017**. *Article ID,* **2017**, *2121878*, 1-16.

[100]   Prior, R.L. Oxygen radical absorbance capacity (ORAC): New horizons in relating dietary antioxidants/bioactives and health benefits. *J. Funct. Foods,* **2015**, *18*, 797-810.
[http://dx.doi.org/10.1016/j.jff.2014.12.018]

[101]   Dorta, E.; Fuentes-Lemus, E.; Aspée, A.; Atala, E.; Speisky, H.; Bridi, R.; Lissi, E.; López-Alarcón, C. The ORAC (Oxygen Radical Absorbance Capacity) index does not reflect the capacity of antioxidants to trap peroxyl radicals. *RSC Advances,* **2015**, *5*(50), 39899-39902.
[http://dx.doi.org/10.1039/C5RA01645B]

[102]   Cheetangdee, N.; Benjakul, S. Antioxidant activities of rice bran protein hydrolysates in bulk oil and oil-in-water emulsion. *J. Sci. Food Agric.,* **2015**, *95*(7), 1461-1468.
[http://dx.doi.org/10.1002/jsfa.6842] [PMID: 25060883]

[103]   Torres-Fuentes, C.; Contreras, M.D.M.; Recio, I.; Alaiz, M.; Vioque, J. Identification and characterization of antioxidant peptides from chickpea protein hydrolysates. *Food Chem.,* **2015**, *180*, 194-202.
[http://dx.doi.org/10.1016/j.foodchem.2015.02.046] [PMID: 25766818]

[104]   Sánchez, D.; Kassan, M.; Contreras, Mdel.M.; Carrón, R.; Recio, I.; Montero, M-J.; Sevilla, M-A. Long-term intake of a milk casein hydrolysate attenuates the development of hypertension and involves cardiovascular benefits. *Pharmacol. Res.,* **2011**, *63*(5), 398-404.
[http://dx.doi.org/10.1016/j.phrs.2011.01.015] [PMID: 21300153]

[105]   Rodríguez-Pérez, C.; Segura-Carretero, A.; Del Mar Contreras, M. Phenolic compounds as natural and multifunctional anti-obesity agents: A review. *Crit. Rev. Food Sci. Nutr.,* **2019**, *59*(8), 1212-1229.
[http://dx.doi.org/10.1080/10408398.2017.1399859] [PMID: 29156939]

[106]   Quirós, A.; Contreras, M.d.M.; Ramos, M.; Amigo, L.; Recio, I. Stability to gastrointestinal enzymes and structure-activity relationship of β-casein-peptides with antihypertensive properties. *Peptides,* **2009**, *30*(10), 1848-1853.
[http://dx.doi.org/10.1016/j.peptides.2009.06.031] [PMID: 19591889]

[107]   Contreras, M.M.; Sevilla, M.A.; Monroy-Ruiz, J.; Amigo, L.; Gómez-Sala, B.; Molina, E.; Ramos, M.;

Recio, I. Food-grade production of an antihypertensive casein hydrolysate and resistance of active peptides to drying and storage. *Int. Dairy J.,* **2011**, *21,* 7.
[http://dx.doi.org/10.1016/j.idairyj.2011.02.004]

[108]   Recio, I.; Contreras, M.M.; Gómez-Sala, B.; Vázquez, C.; Fernández-Escribano, M.; del Campo, R. effect of a casein hydrolysate containing novel peptides in hypertensive subjects. *Ann. Nutr. Metab.,* **2011**, *58*, 16-17.

[109]   Drucker, D.J. Dipeptidyl peptidase-4 inhibition and the treatment of type 2 diabetes. *Bench to Clin. Symp.,* **2007**, *30*(6), 1335- 1343.
[http://dx.doi.org/https://doi.org/10.2337/dc07-0228.D.J.D]

[110]   Lee, N.K.; Paik, H.D. Status, antimicrobial mechanism, and regulation of natural preservatives in livestock food systems. *Han-gug Chugsan Sigpum Hag-hoeji,* **2016**, *36*(4), 547-557.
[http://dx.doi.org/10.5851/kosfa.2016.36.4.547] [PMID: 27621697]

[111]   Tan, Y.N.; Ayob, M.K.; Wan Yaacob, W.A.; Yaacob, W. Purification and characterisation of antibacterial peptide-containing compound derived from palm kernel cake. *Food Chem.,* **2013**, *136*(1), 279-284.
[http://dx.doi.org/10.1016/j.foodchem.2012.08.012] [PMID: 23017424]

[112]   Aider, M.; Barbana, C. Canola proteins: composition, extraction, functional properties, applications as a food ingredient and allergenicity - A practical and critical review. *Trends Food Sci. Technol.,* **2011**, *22*(1), 21-39.
[http://dx.doi.org/10.1016/j.tifs.2010.11.002]

[113]   Burlando, B.; Cornara, L. Therapeutic properties of rice constituents and derivatives (*Oryza sativa* L.): A review update. *Trends Food Sci. Technol.,* **2014**, *40*(1), 82-98.
[http://dx.doi.org/10.1016/j.tifs.2014.08.002]

[114]   Regulation (EU) 2015/2283 of the European Parliament and of the Council of 25 November 2015 on novel foods, amending regulation (EU) No 1169/2011 of the European Parliament and of the Council and Repealing Regulation (EC) No 258/97 of the European Parliam. *Off. J. Eur. Union,* **2015**, *L327*(258), 1-22.

[115]   Karthikeyan, O.P.; Mehariya, S.; Chung Wong, J.W. Bio-refining of food waste for fuel and value products. *Energy Procedia,* **2017**, *136*, 14-21.
[http://dx.doi.org/10.1016/j.egypro.2017.10.253]

[116]   European Commision. *A Sustainable Bioeconomy for Europe: Strengthening the Connection between Economy*; Society and the Environment, **2018**.

[117]   Haschek, W. M.; Rousseaux, C. G.; Wallig, M. A. Kidney and lower urinary tract. *Fundam. Toxicol. Pathol.,* **2010**, 261- 318.
[http://dx.doi.org/https://doi.org/10.1016/b978-0-12-370469-6.00011-8]

[118]   Getachew, A.T.; Chun, B.S. Influence of pretreatment and modifiers on subcritical water liquefaction of spent coffee grounds: A green waste valorization approach. *J. Clean. Prod.,* **2017**, *142*, 3719-3727.
[http://dx.doi.org/10.1016/j.jclepro.2016.10.096]

[119]   Chambi, H.N.M.; Lacerda, R.S.; Makishi, G.L.A.; Bittante, A.M.Q.B.; Gomide, C.A.; Sobral, P.J.A. Protein extracted from castor bean (*Ricinus communis* L.) cake in high pH results in films with improved physical properties. *Ind. Crops Prod.,* **2014**, *61*, 217-224.
[http://dx.doi.org/10.1016/j.indcrop.2014.07.009]

[120]   Anadón, A.; Martínez, M.A.; Ares, I.; Ramos, E.; Martínez-Larrañaga, M.R.; Contreras, M.M.; Ramos, M.; Recio, I. Acute and repeated dose (4 weeks) oral toxicity studies of two antihypertensive peptides, RYLGY and AYFYPEL, that correspond to fragments (90-94) and (143-149) from α(s1)-casein. *Food Chem. Toxicol.,* **2010**, *48*(7), 1836-1845.
[http://dx.doi.org/10.1016/j.fct.2010.04.016] [PMID: 20398720]

CHAPTER 2

New Developments in the Quinolone Class of Antibacterial Drugs

Neslihan Demirbas[*] and **Ahmet Demirbas**

Karadeniz Technical University, Department of Chemistry, 61080 Trabzon, Turkey

Abstract: The increasing drug resistance and the insufficiency of the newly developing antibiotics constitute a serious and growing health threat in the world. Especially Gram (-) bacteria acquire genetic material encoding antibiotic resistance by multiple mechanisms. Development of novel antibacterial agents with little tendency to bacterial resistance is, therefore, an important and challenging topic in the medicinal chemistry, and synthetic organic chemistry is an indispensable part of the design and synthesis of efficient antibacterial drug candidates. Among the broad-spectrum antibiotics, fluoroquinolones constitute the most attractive drugs in the anti-infective chemotherapy field. These antibiotics target the bacterial type II topoisomerase enzymes (DNA gyrase and topoisomerase IV) which are essential enzymes involved in bacterial cell growth and division. Since their advent, they were widely applied to treat infections. Unfortunately, most of them suffered from the resistance problem by mutations in the bacterial targets due to their wide use. Recently, the synthetic organic and medicinal chemists focused their research on the design of new fluoroquinolones with improved features by molecular hybridization technique. One of the most promising approaches aiming to combat resistant pathogens is the design and synthesis of new hybrid molecules in which different pharmacophore groups with different modes of action are joined together using a flexible linker. This strategy supplies a way to improve traditional drug combination therapies simplifying optimization of the pharmacokinetics/pharmacodynamic (PK/PD) profile, efficacy at both targets is usually synergistic.

Keywords: Aminoglycoside, Drug resistance, Flavonoid, β-Lactam, Macrocyclic, Molecular hybridization, Oxazolidinone, Pyrazole, Pyrazine, Pyrimidine, Quinolone, Triazole.

INTRODUCTION

In recent years, the growing incidence of virulent bacterial resistance towards the present antibacterial agents has become the most serious clinical and socio-

[*] **Corresponding author Neslihan Demirbas:** Karadeniz Technical University, Department of Chemistry, 61080 Trabzon, Turkey; Tel/Fax: +90 462 3774252; E-mail: neslihan@ktu.edu.tr

Atta-ur-Rahman (Ed.)

economic problem worldwide [1 - 3]. Although, The World Health Organization, has described the antibiotics as "miracle weapons giving an opportunity to combat with infectious diseases", a large majority of clinically effective drugs actively used to treat bacterial infections have become less effective due to the increasing antimicrobial resistance [4 - 9]. Moreover, the treatment of infectious diseases is more difficult in immunodeficient patients, such as those infected with tuberculosis, HIV *etc* [9]. Multidrug resistant Gram (+) pathogens, such as methicillin-resistant Staphylococcus aureus (MRSA) and *Staphylococcus epidermis* (MRSE), vancomycin-resistant *Enterococci* (VRE), cephalosporin resistant *Streptococcus pneumoniae* are leading significant morbidity and mortality of the infected patients [10 - 12]. Another pathogen, penicillin resistant *S. pneumoniae* has been reported to cause approximately 3 million deaths each year worldwide because of pneumonia, meningitis and sepsis, which are responsible for serious upper airway infections, such as sinusitis and otitis media [13 - 16].

Microorganisms develop resistance to drugs *via* various mechanisms, such as overexpression of drug efflux transporters, like multidrug and toxic compound extrusion (MATE) transporters [17], changes in the target sites of antibiotics [18], optimization of the enzyme (such as β-lactamase) activity resulting in inactivation of antibiotics [19], spontaneous chromosomal mutations [20], and horizontal transfer of genetic elements [21]. Inhibition of the activity of drug efflux transporters appears to be an encouraging strategy for renovating the activity of a drug that is the substrate of these efflux pumps [22].

Keeping all this in mind, it is clearly seen that the development of wholly novel drug discovery methodologies and the optimization of available antibacterial agents have become a crucial and challenging task for the effective treatment of bacterial infections. However, the development of completely new antibacterials suitable for therapeutic applications has not been as successful as expected, and despite a tenfold increase in spending for Research-Development studies in the pharmaceutical industry, the number of leader molecules has remained nearly stable.

To improve the therapeutic profile of the existing drugs by several manipulations in their structures or to design their novel analogs has become one of the most promising strategies for the development of new antibacterial drugs. This strategy has been widely admitted since it does not entail to discover novel scaffolds or validation of new biological targets, which has been accepted as an extremely difficult and time-consuming procedure [27].

In recent years, in order to overcome the "drug resistance nightmare", the concept of "molecular hybridization" based on the combination of structural features of two or more drug fragments having different modes of action has emerged as an attractive strategy. These new hybrid compounds with improved affinity and efficacy have been proved to be capable of inhibiting two or more conventional targets simultaneously, and this multiple target strategy has led to discover a number of bioactive hybrid molecules [28 - 31].

In recent drug development programs, 4-quinolone-3-carboxylic acid scaffold has been used as one of the most frequently encountered privilege frameworks having potent and broad spectrum activity [32]. Since the introduction of nalidixic acid (the first generation, the prototype 4-quinolone antibiotics) for the treatment of urinary tract infections in humans in 1962, the class of quinolone antibacterials has played an important role saving countless millions of lives in the chemotherapy of bacterial infections [33 - 35]. Their preferable properties including well tolerability with excellent safety profile, favorable pharmacokinetic characteristics, broad antibacterial spectrum and good treatment effectiveness have made quinolones an important class of synthetic antibacterial agents [36, 37]. This class of antibacterials displays direct inhibition activity on the DNA synthesis by binding to the enzyme DNA complex, they stabilize DNA strand breaks created by DNA gyrase and topoisomerase IV [38]. Gyrase is responsible for introducing negative supercoils in DNA and relieving torsional stress expected to accumulate ahead of transcription and replication complexes. Topoisomerase IV provides a potent decatenating activity. Both gyrase and topoisomerase IV are essential enzymes and therefore the compounds that block bacterial growth by inhibiting them are accepted as potential chemotherapeutics [39].

CLASSIFICATION, SYNTHESIS AND STRUCTURAL REQUIRE-MENTS OF QUINOLONE ANTIBACTERIALS

Until today, four generations of quinolone class antibacterial drugs have been developed.

First Generation

The first generation includes: nalidixic acid, oxolinic acid, pipemidic acid, cinoxacin and rosoxacin which have shown only weak to moderate activity against Gram (-) bacteria, and therefore is rarely used today.

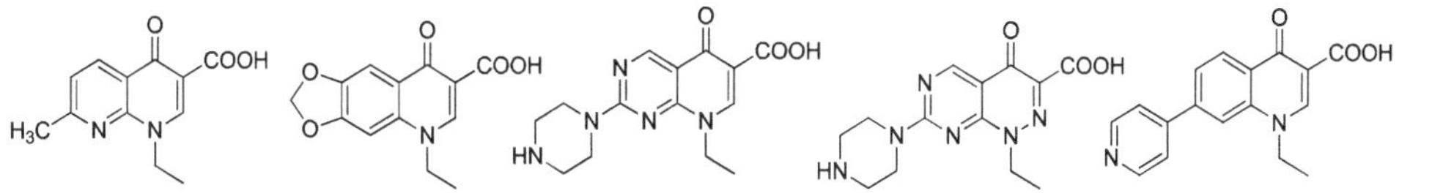

Nalidixic acid   **Oxolinik acid**   **Pipemidic acid**   **Cinnoxacin**   **Rosoxacin**

Second Generation

The second generation, defined as fluoroquinolones (**FQs**), includes one or more fluorine atoms. This group contains norfloxacin, pefloxacin, enoxacin, ofloxacin, ciprofloxacin, lomefloxacin, fleroxacin, rufloxacin, nadifloxacin, exhibiting high activity on Gram (-) bacteria, besides considerable activity on Gram (+) bacteria.

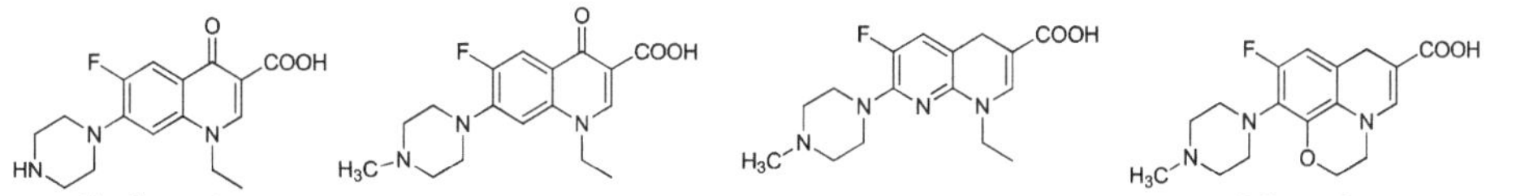

Norfloxacin   **Pefloxacin**   **Enoxacin**   **Ofloxacin**

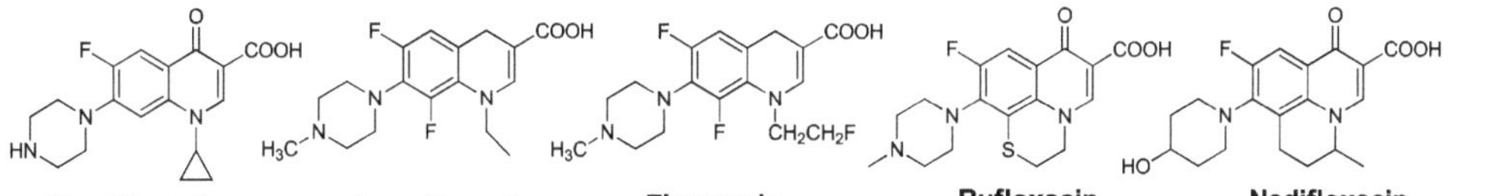

Ciprofloxacin   **Lomefloxacin**   **Fleroxacin**   **Rufloxacin**   **Nadifloxacin**

Third Generation

The third generation consisting of levofloxacin, pazufloxacin, temafloxacin, tosufloxacin, sparfloxacin, grepafloxacin and balofloxacin is also found to be active towards streptococci.

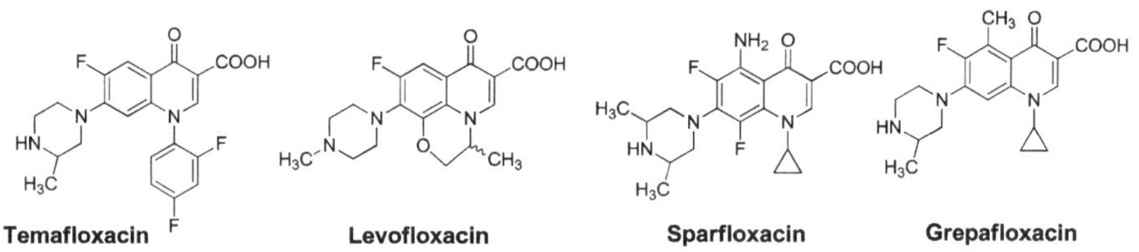

Temafloxacin   **Levofloxacin**   **Sparfloxacin**   **Grepafloxacin**

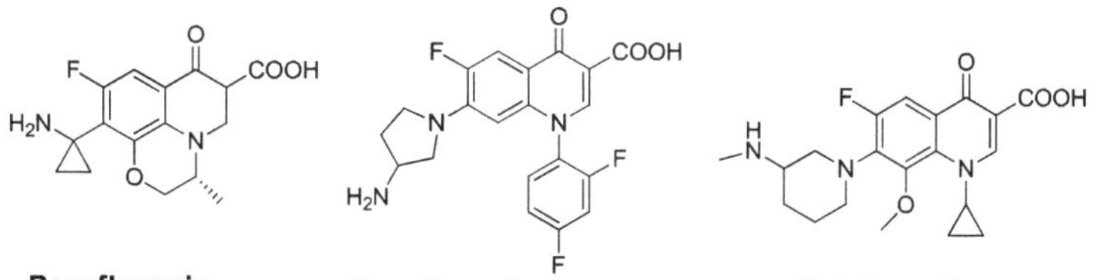

Pazufloxacin   **Tosufloxacin**   **Balofloxacin**

Fourth Generation

The fourth generation consists of gatifloxacin, gemifloxacin, sitafiloxacin, clinafloxacin, trovafloxacin, moxifloxacin, alatrofloxacin, delafloxacin, besifloxacin, garenoxacin, finafloxacin, and is characterized by good activity against Gram (-), Gram (+) and anaerobic bacteria, and acts at both DNA gyrase and topoisomerase IV, thus slowly developing the resistance [40].

Gatifloxacin   Gemifloxacin   Sitafloxacin   Clinafloxacin

Trovafloxacin   Moxifloxacin   Alatrofloxacin

Delafloxacin   Besifloxacin   Garenoxacin   Finafloxacin

Synthesis of Some Quinolone Drugs

Rosoxacin

Norfloxacin, Ciprofloxacin, Fleroxacin

The most common changes in the structural modifications of the antimicrobial quinolones are schematized in Fig. (**2**) [40].

The numbering of 4-quinolone-3-carboxilic acid skeleton is presented in Fig. (**1**).

Fig. (1). The numbering of 4-quinolone-3-carboxylic acids.

However, the increasing resistance to the quinolone class antibacterial drugs hasemerged as a growing health problem in the community and medical settings, and this event emphasizes the urgent need for safe and affordable antibacterial drugs with a lesser tendency to resistance [42]. The studies devoted to this target have focused on two strategies mainly. Although to define completely new targets having a novel mechanism of action is the ideal one, this approach has been less successful, because, almost all 4-quinolones exhibited the same activity mechanism containing the inhibition of DNA gyrase and topoisomerase IV. The other approach contains the synthesis of new analogs or modification of the existing drugs since this strategy does not require target validation studies and supply an easy way to generate new agents with limited or no cross-resistance to currently used drugs. Due to this reason, the second methodology has been successfully applied for the development of new series of quinolone class antibacterial agents [43]. In 4-quinolones, the major domains are the positions *N*-1, *C*-5, *C*-6, *C*-7 and *C*-8 for chemical modifications. 3-Oxo-4-carboxylic acid core has been accepted as essential for interaction with DNA bases over the formation of H-bonding. Relatively small and lipophilic substituents are favored for *N*-1 and C-8 positions [44]. The optimal substituents at C-5 and C-6 are -NH$_2$ and -F, respectively, for inhibition on Gram (+) bacteria. The position C-7 represents the site suitable for further modifications, and the nature of the substituent at C-7 has a remarkable effect on the potency, spectrum, safety and pharmacokinetics of newly generated compounds [42 - 45].

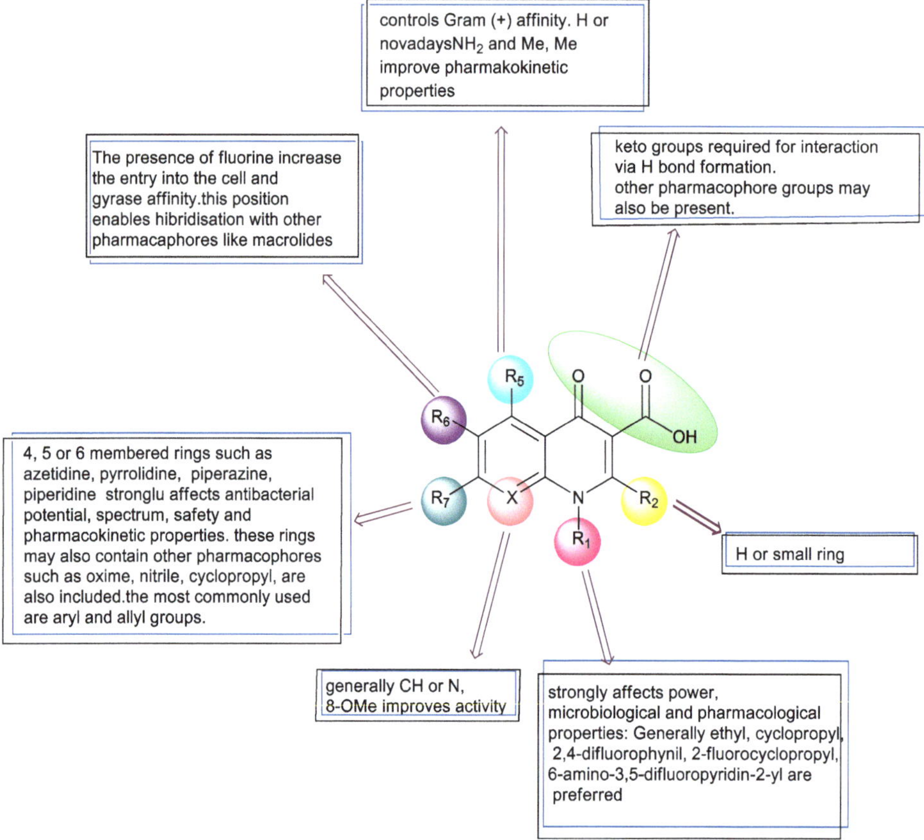

Fig. (2). The most common changes in the structural modifications of the antimicrobial quinolones.

The *N*-1 substituent has been accepted to influence the potency, pharmacodynamic and pharmacokinetic properties [40], and various substituents have been introduced to this position. Ethyl and cyclopropyl are the common substituents for first generation and second generation quinolones, while 2,4-difluorophenyl is highly utilized groups for third and fourth generation quinolones. Among them, 5-amino-2,4-difluorophenyl group has been reported to be more potent than the others [46]. Previous studies have shown that the large groups in C-7 positions do not induce undesirable effects on the penetration capacity of the drug through bacterial membranes, and at the same time, this group modifies the strength of action significantly [47].

So far, much effort has been devoted to the effect of the structural modifications at the C-7 position in quinolones on antibacterial activity. The C-7 substituents strongly affect antibacterial potency, spectrum, bioavailability, solubility, safety and pharmacokinetic properties, and this position is regarded as the most adaptable site for chemical change [40]. *N*-Containing five or six-membered heterocycles such as pyridine [40, 48], thiazolidine [49], pyrrolidine [50 - 52], piperidine [53], piperazine [54 - 58] have been introduced into this position of quinolones. Other groups located on the C-7 position of quinolone skeleton are nitrothiophene [59], nitroimidazole [60], nitrofuran [61], 1,3,4-thiadiazole [62, 63], isatin [64], bromothiophene [65], acyl groups [66] triazole [67, 68] and other substituents [69 - 73]. Until today, the efforts devoted to discover new quinolone derivatives with activity on drug-resistant bacteria have resulted in countless new compounds.

The structural modifications in the quinolone core at the C-3 position are relatively lesser than the other positions. Some studies have unclosed that it might be attributed to the specific mutations of amino acids in gyrase and topoisomerase IV that disrupted the carbonyl group of C-3 position, water and metal ion bridge and weakened interactions between quinolones and these enzymes [67]. Moreover, some side effects, such as loss of metal ions and gastrointestinal disorders caused by quinolones have been ascribed to the carbonyl group at C-3 position.

Therefore, the structural modification at the C-3 position in the quinolone skeleton is an encouraging strategy to obtain new antibacterials with improved properties and to overcome the drug-resistance.

It has been known that the presence of fluorine atom as a highly electronegative center increases lipophilicity, and can lead to change in the physico-chemical properties of a molecule. This can cause an increase in the stability, decrease in toxicity and improve the antibacterial activity, as a result, the overall therapeutic efficacy can enhance [74]. The introduction of trifluoromethyl group in quinolone structure can also cause improved therapeutic performance. But in this case, the hydrophobicity of the molecule is higher than the case of the presence of fluorine, because trifluoromethyl group increases the volume causing a larger molecule [75]. The trifluoromethyl substituent has been found in the structures of several bioactive compounds, such as herbicides, fungicides, analgesic and antipyretic [75]. Panda *et al.* designed the synthesis of new potential antimicrobial agents consisting of two biodynamic scaffolds, quinolone with C-2 trifluoromethyl substituent and heterocyclic amine, which is connected *via* a spacer (Scheme **3**) [75].

The co-application of bacterial efflux pump inhibitors (EPIs) with antibacterial drugs has been maintained by scientific and commercial laboratories to overcome efflux-mediated resistance to medically used antibacterial agents [76 - 79]. These attempts have shown that the inhibition of an efflux pump system leading to efflux of an antibiotic can re-establish or strengthen the activity of that drug [80]. The inhibition of efflux pumps also prevents the emergence of higher level resistance mechanisms, such as target mutations that occur as a result of up-regulated efflux pumps. During the last decade, a large number of efflux pump system inhibitors with different structures have been discovered by natural sources, screening of compound libraries, and re-evaluation of the known therapeutics [76]. However, most of the EPIs have limited inhibition capacity on efflux systems, and this is attributed to the presence of diversity and the number of pump types. This has been accepted as an obstacle to discover new efflux pump inhibitors finding wide clinical use in medical therapy. The co-implementation of EPIs capable to inhibit two different pump types has also exhibit synergism, when used together with certain substrates in Gram-negative organisms. It is known that the integration of FQs with EPIs or the introduction of EFs units to FQ structure leads to expand the spectrum of Eps. German and coworkers have synthesized new fluoroquinolones *via* the integration of ciprofloxacin and ofloxacin with structural motifs of two class of pump inhibitors (Fig. **3**) at the C-7 (Schemes **4-6**) [76].

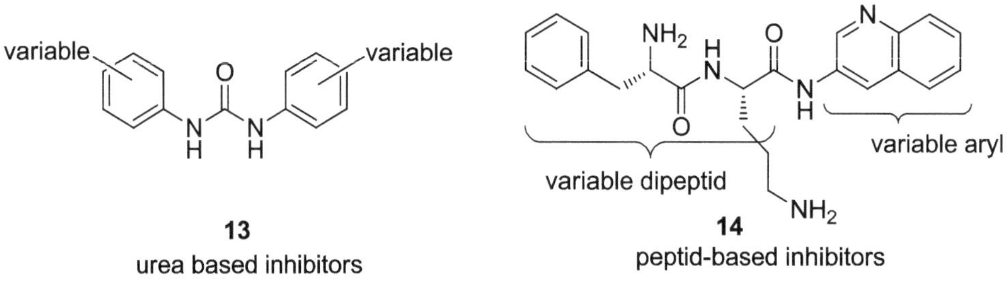

Fig. (3). Schematic representation of two classes of pump inhibitors.

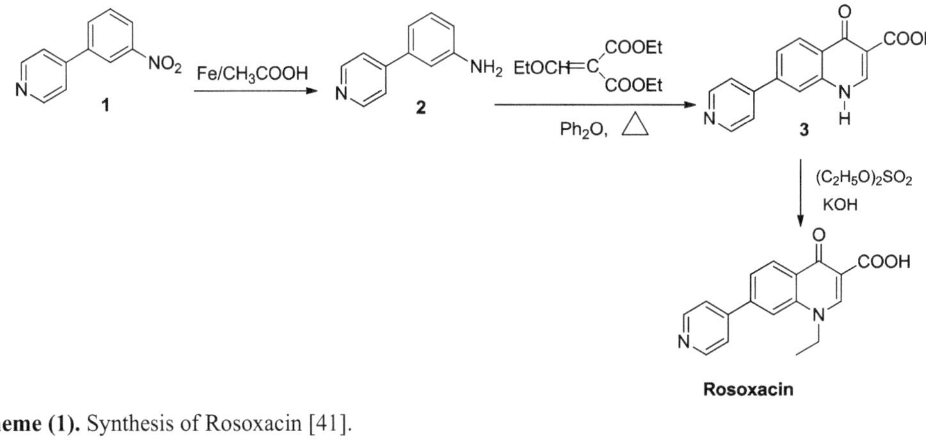

Scheme (1). Synthesis of Rosoxacin [41].

Scheme (2). Synthesis of Norfloxacin and Ciprofloxacin.

Scheme (3). Synthetic pathway for the preparation of new quinolones consisting of two biodynamic scaffolds.

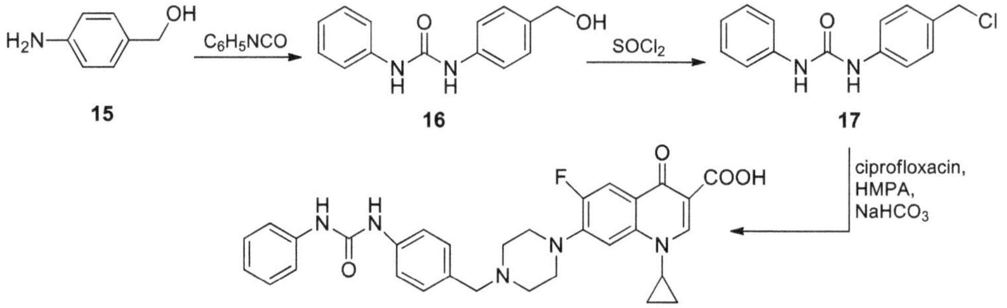

Scheme (4). Synthesis of ciprofloxacin derivative bearing a bisaryl urea EPI at the C-7 position.

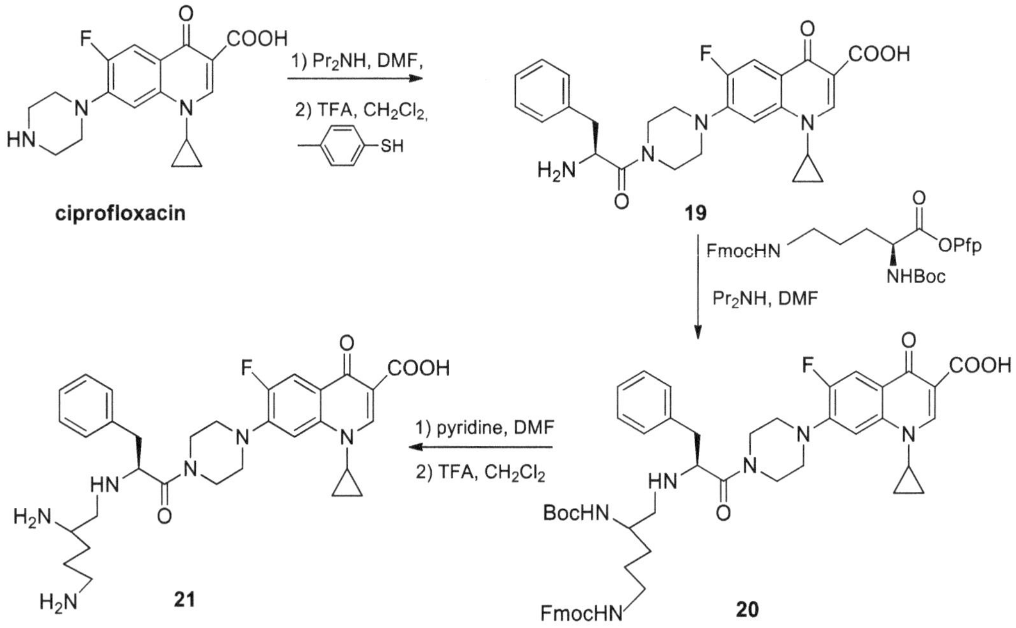

Scheme (5). Synthesis of ciprofloxacin derivative bearing the L-ornithine-L-phenylalanine C-7 dipeptide.

Scheme (6). Synthesis of ciprofloxacin derivative bearing the L-phenylalanine-L-ornithine C-7 dipeptide.

Allicin, a natural electrophilic (δ^+) thiosulfinate from Allium sativum (garlic), has been reported to inhibit biofilm formation by bacteria and display antibacterial activity. *N*-Thiolated fluoroquinolones have been synthesized by Sheppard and Long with the aim to introduce allicin unit to fluoroquinolone scaffold (Scheme **7**) [81].

THE PROGRESS OF HYBRIDIZATIONS IN QUINOLONES

Hybridization with Azetidines

The compounds reported by Itoh and coworkers, which contain 2,4-difluoro-5-aminopyridine and 3-aminoazetidine substituents at *N*-1 and C-7, respectively, may be regarded as new analogs of delafloxacin (Scheme **7**) [45]. According to the results obtained by Itoh and coworkers, the *N*-1 substituent with the amino group has led to the reduction of the phototoxicity, even when halogen atoms that induce phototoxicity have been introduced at the C-8 position. Furthermore, a four-membered heterocyclic ring, such as aminoazetidine, at C-7 position caused to a higher antibacterial activity than a five-membered ring, such as aminopyrrolidine in combination with the presence of difluoro-aminopyridine as *N*-1 substituent; and the presence of methyl group at C-8 was found to be effective for reducing the cytotoxicity without the loss of *in vitro* activity [73]. Moreover, Frigola and coworkers have reported that the introduction of an alkyl group to aminoazetidine moiety located at C-7 improves the pharmacokinetic properties, but the length of the alkyl group is not clear (Scheme **8**) [82 - 84].

R= methyl, ethyl, *n*-propyl, *i*-propyl, *n*-butyl, *s*-butyl, 3-methyl-2-butyl, *n*-hexyl

Scheme (7). Synthesis of *N*-thiolated fluoroquinolones.

Scheme (8). Synthesis of 3-aminoazetidine fluoroquinolones.

Ikee *et al.* have been synthesized new C7-azetidinyl-sulfenyl and C7-azetidinyl fluoroquinolones with antibacterial activity, in order to determine the effect of azetidine ring on biological activity, and most of the new compounds have exhibited good inhibition potency towards Gram (+) bacteria (Schemes **9, 10**) [85].

Scheme (9). Synthesized new C7-azetidinyl-sulfenyl fluoroquinolones.

a: R= -C$_6$H$_5$; R'= H; **b:** R= -C$_6$H$_4$OMe(*4*-), R'= H; **c:** R= -C$_6$H$_4$Me(*4*-), R'= H

d: R= -C$_6$H$_4$Cl(*4*-); **e:** R= -C$_6$H$_4$COOEt(*4*-), R'= H; **f:** R= R'= -CH$_2$C$_6$H$_5$

Scheme (10). Synthesis of C7-azetidinyl fluoroquinolones.

Hybridization with Triazole

1,2,3-Triazole ring system is a very well-recognized pharmacophore [36 - 38] which is derived from Huisgen's 1,3-dipolar cycloaddition of azides and alkynes by copper-catalyzed click reaction [39].

a: R= -$C_6H_3F_2$(*2,4*-); b: R= -C_6H_4Cl(*4*-); c: R= -C_6H_3FCl(*3,4*-); d: R= -C_6H_4Br(*4*-);

e: R= -C_6H_3ClF(*2,4*-); f: R= -C_6H_3ClMe(*2,4*-); g: R= -$C_6H_3Cl_2$(*2,4*-); h: R= -$C_6H_4OCF_3$(*4*-);

i: R= -$C_6H_3Cl_2$(*2,3*-); j: R= -$C_6H_4NO_2$(*4*-); k: R= -C_6H_4Cl(*2*-); l: R= -C_6H_4Cl(*2*-)

Scheme (11). Synthetic pathway leading the formation of 1,2,3-triazole-ciprofloxacin conjugates.

This nitrogen-containing heterocycle is renowned among U.S. FDA approved pharmaceuticals [40]. The moiety is an attractive connecting unit, since it is stable to metabolic degradation, oxidative/reductive conditions and actively participates in binding to biomolecular targets and improves its solubility by hydrogen bonding and dipole interactions [41, 42]. Inspired by the observations mentioned above, Kant *et al.* have designed a series of 1,2,3-triazole-ciprofloxacin conjugates as potent antibacterial agents (Scheme **11**) [37]. The optimization of the potency of the synthesized compounds has also been performed by these researchers.

1,2,4-Triazole ring has been accepted as a convenient chief molecule, leading to the production of new potential bioactive structures. A number of compounds containing this heterocycle have exhibited a wide variety of therapeutically important activities, such as antibacterial [86, 87], anti-inflammatory [88], CNS depressant [89], anti-tubercular [90] and antitumoral [91 - 93] *etc.* 1,2,4-Triazole nucleus constitutes a structural element of azole class antifungal drugs including voriconazole, itraconazole, fluconazole, miconazole, econazole, isoconazole, posaconazole *etc* (Fig. **4**) [94 - 101].

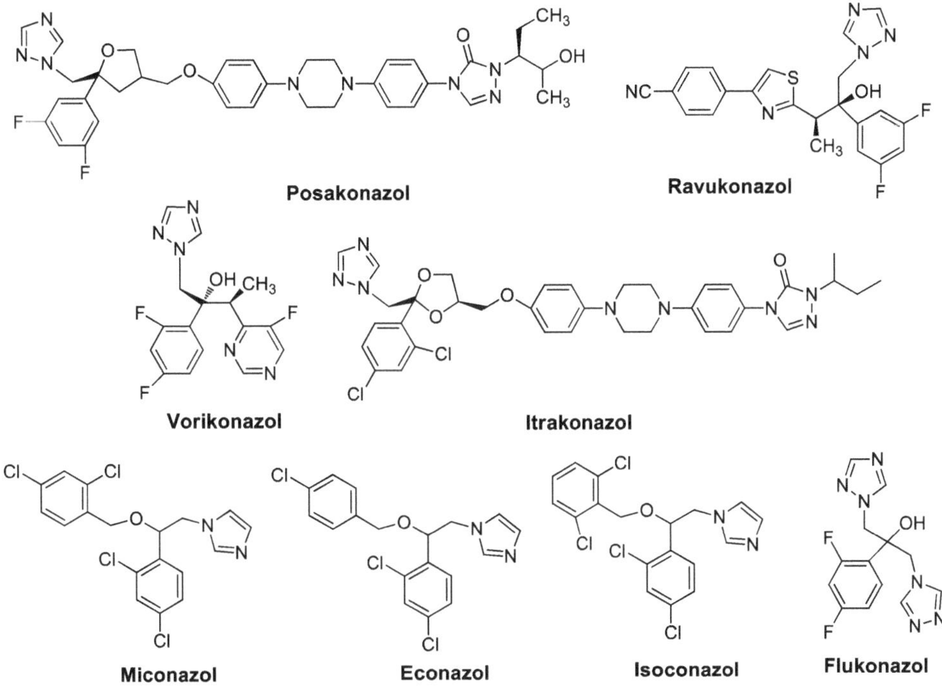

Fig. (4). Some known azole class antifungal drugs.

Several other compounds including 1,2,4-triazole unit have also been well known as drugs, *e.g.*, triazolam, alprazolam, estazolam, trazodone, nefazodone, rizatriptan and ribavirin (Fig. **5**).

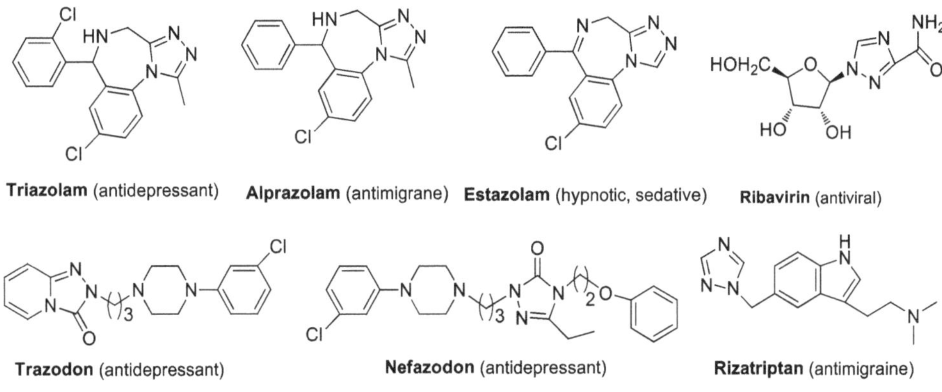

Fig. (5). 1,2,4-Triazole containing drugs.

Other drugs vorozole, letrozole and anastrozole which have been known as aromatase inhibitors and used especially for clinical therapy of breast cancer are triazole derivatives (Fig. **6**) [97].

Fig. (6). Vorozole, letrozole and Anastrozole.

Their nuclophilic centers make triazoles good starting compounds leading to the formation of several *N*- and *S*-containing heterocyclic rings.

The biological activity exhibited by quinolone hybrids is not only antibacterial activity, but they also exhibit anti-urease and/or antioxidant activity [102 - 105]. Some fluoroquinolone-triazole hybrids (**47a-h**) have been reported by Mermer *et al.* as antibacterial and antioxidant agents (Fig. **7**) [105].

a, b: R= -C$_6$H$_5$, **c, d**: R= -C$_2$H$_5$; **e-h**: R= -CH$_2$C$_6$H$_5$; **a-d, g, h**: X=S,

e, f: X= O; **a,c,e,g**: R'= -C$_2$H$_5$; **b, d, f, h**: R'= -C$_6$H$_5$, R'=

Fig. (7). Fluoroquinolone-triazole hybrids with antibacterial and antioxidant activity.

Multicomponent reactions (MCRs) have provided an easy and useful procedure to design complex molecules with biological activity. In comparison with conventional organic reactions, MCRs have some excellent properties such as high conversion rate, minimal reaction time, structurally complexity and more importantly they provide an ecofriendly way to access biologically active or synthetically important products. Among MCRs, Mannich reaction being a one pot-three component condensation, leads to the construction of β-aminoalkylated compounds having biological activity [105, 106]. Plech and coworkers have prepared some ciprofloxacin-triazole hybrids having antibacterial activity even on resistant bacteria, through Mannich condensation (Scheme **12**) [107, 108].

a-k: R= 3-Cl-C_6H_4; **l-o**: R= 3-OH-C_6H_4; **p**: R= H; **q**: R= CH_3; **r**: R= n-C_3H_7; **a**: R'= 4-$OCH_3C_6H_4$; **b**: R': 4-Br-C_6H_4;

c, m, p-r: R'= 3-I-C_6H_4; **d**: R'= 4-I-C_6H_4; **e, n**: R'= 2,4-$F_2C_6H_3$; **f, n**: R'= 2,4-$Cl_2C_6H_3$; **g, o**: R'= 3,4-$Cl_2C_6H_3$;

h: R'= CH_3; **i**: R'= C_2H_5; **j**: R'= n-C_7H_7; **k**: R'= n-C_6H_{13}; **l**: R'= 3-$NO_2C_6H_4$; **m; p-r**: R'= 3-I-C_6H_4

Scheme (12). Synthesis of some ciprofloxacin-triazole hybrids.

In order to design more active and safe quinolone antibacterials, Wang *et al.* synthesized clinafloxacin-fluconazole hybrids with both antibacterial and antifungal activities (Scheme **13**) [109]. Clinafloxacin belongs to a fluoroquinolone class of antibacterials that are more active than the other fluoroquinolone antibiotics with a broader activity towards most of the Gram-positive, Gram-negative and anaerobic bacteria. On the other hand, fluconazole, that has a triazole nucleus in its structure (Fig. **4**), has been used as one of the most important antifungal drugs. But, the drug resistance limits the effective use of fluconazole. Moreover, fluconazole could not be used successfully in inhibiting invasive aspergillosis [109]. Therefore, the structural modifications of fluconazole have primarily focused on the side chain of this drug by the replacement with other pharmacophores which lead to increase in the binding ability between the drug and fungus. In order to design more active and safe clinafloxacin-based triazole hybrids, Wang *et al.* have combined the antifungal fluconazole with antibacterial clinafloxacin [109]. It has been accepted that 1,2,4-triazole ring can enforce interactions with DNA, enzymes and receptors *via* hydrogen bonds, coordination, ion-dipole, cation-π, π–π stacking, hydrophobic effect and/or van der Waals force *etc.* as a result of its extraordinary five-membered aromatic structure containing three nitrogen atoms [109]. This means that the integration of clinafloxacin with 1,2,4-triazole unit might cause to diverse interactions with various targets, which may correspond to different mechanism of action than the parent compounds, contributing to overcome the drug resistance problem. The results of such a combination not only improve their physicochemical property and binding affinity, and thereby increase biological activity and widen the activity spectrum, but also increase the water solubility of the new hybrids *via* H-bond formation by nitrogen atoms in triazole (Scheme **13**) [109].

a: $X_1=X_2=$ H, $X_3=$ CH$_3$; **b**: $X_1=X_2=$ H, $X_3=$ Cl; **c**: $X_1=X_2=$ H, $X_3=$ F; **d**: $X_1=$ H, $X_2=X_3=$ Cl;

e: $X_1=$ H, $X_2=X_3=$ F; **f**: $X_1=X_3=$ Cl, $X_2=$H; **g**: $X_1=X_3=$ F, $X_2=$ H

Scheme (13). Combination of fluconazole with clinafloxacin.

Gao and coworkers have fused the 1,2,4-triazole ring to the *N*-1 and C-8 positions in the common fluoroquinolone intermediate (**55**) to give the tricycle-fused quinoline-3-carboxylic acid derivatives, as antibacterial compounds (Scheme **14**) [110].

a: dimethyl amine, **b**: diethyl amine, **c**: piperidine, **d**: morpholine,

e: *N*-methyl piperazine, **f**: *N*-ethyl piperazine, **g**: 2-methyl piperazine,

h: pyrrolidine

Scheme (14). Synthesis of 1,2,4-triazole-fused quinoline-3-carboxylic acids.

Fig. (**8**) summarizes the structures of dihydrotriazolo[4,5-h]quinolone-carboxylic esters and their acids (**59**) which have been reported by Carta *et al.* as a new class of quinolones with potent anti- mycobacterial activity, while their isomers **55** and **56** were completely inactive on mycobacteria [111].

Fig. (8). Dihydrotriazolo[4,5-h]quinolone-carboxylic esters and acids.

Aggarwal *et al.* have reported new quinolone hybrids possessing antimycobacterial activity by the introduction of triazole, triazolothiadiazole or triazolothiadiazine ring to the C-3 position of nalidixic acid (Scheme **15**) [112].

Scheme (15). Synthesis of nalidixic acid hybrids containing triazole, triazolothiadiazole or triazolothiadiazine core.

Novel antibacterial nalidixic acid-benzotriazole and oxolinic acid-benzotriazole amides at C-3 have been reported by Panda *et al* (Scheme **16**) [27].

Scheme (16). Preparation of nalidixic acid-benzotriazole and oxolinic acid-benzotriazole hybrids.

The reports of Hu and coworkers have shown that the introduction of 1,2,4-triazole-5(4*H*)-thione nucleus to quinolone skeleton has greatly affected the *in vitro* antibacterial activity against both Gram-negative and Gram-positive bacteria and most of them are more active than the parent quinolones. Phenyl ring with an electron releasing group is mandatory for excellent activity. On the other hand, phenyl group at the *N*-4 position of the triazole ring is not vital, even shorter alkyl groups have been preferred for high activity. The replacement of *N*-3-substituent by oxygen does not decrease the activity, while the elimination of the thione function is detrimental for activity (Figs. **9**, **10**) [50, 106, 107].

Fig. (9). Building requirements in 4-quinolone-triazole hybrids.

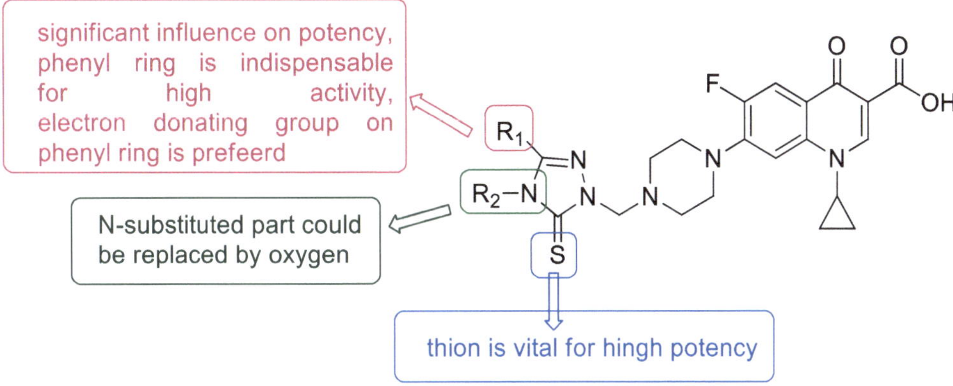

Fig. (10). Building requirements in the triazole part of 4-quinolone-triazole hybrids.

Further studies have suggested that improved activity compared with parent quinolones can be attributed to these:

1. Enhanced permeation profile;
2. Novel mechanism of action which is not incorporated in topoisomerases inhibition
3. Exclusion from bacterial efflux pumps.

Mentese *et al.* have designed some ciprofloxacin hybrids with antibacterial activity *via* the introduction substituted 1,2,4-triazole unit into 4-position of piperazine ring on ciprofloxacin (Scheme **17**) [99].

Based on the molecular hybridization concept, novel molecules in which thiomorpholine nucleus has been incorporated into fluoroquinolone skeleton *via* an azole linker which hasbeen synthesized by Demirci *et al.* These hybrids have shown antibacterial, anti-α-glucosidase, antilipase, antiurease and/or antioxidant activities (Scheme **18**). In this study, 1,2,4-triazole or 1,3,4-oxadiazole nucleus has been introduced to fluoroquinolone scaffold *via* a Manncih type condensation in the microwave prompted media [100].

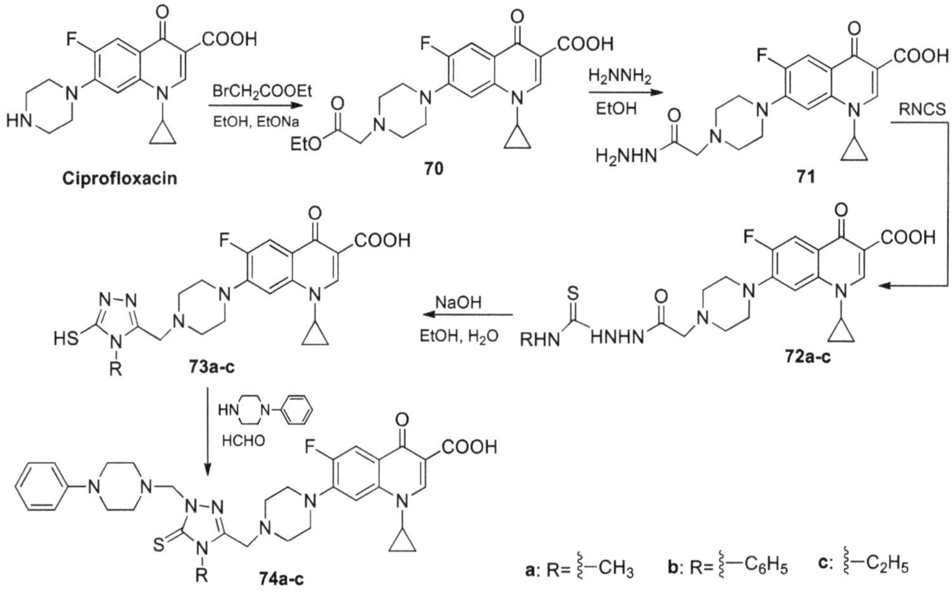

Scheme (17). Preparation of ciprofloxacin-triazole-phenylpiperazine hybrids through mannich reaction.

Thiomorpholine moiety has extensively been found in the structures of biologically active products because of their favorable lipophilic and hydrophilic properties [100]. For instance, sutezolid (Fig. **11**), that belongs to oxazolidinone class antibiotics in phase II clinical trials, constitutes one of the most commonly known examples [100].

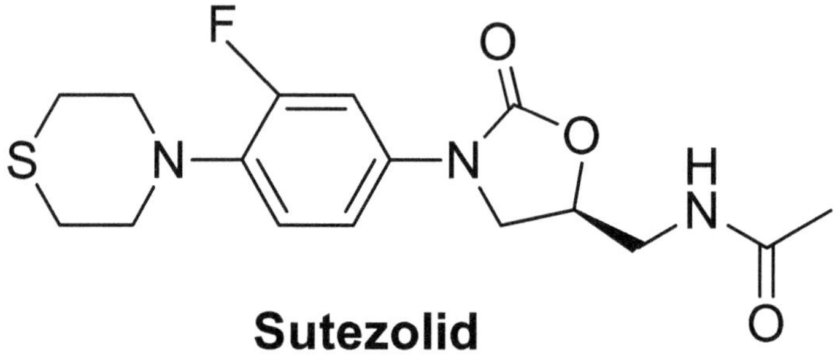

Fig. (11). Sutezolid.

80a: R= -CH$_2$C$_6$H$_5$, X= S; **80b**: R= -C$_6$H$_5$, X=O; **81a**: R= -CH$_2$C$_6$H$_5$, X= S, R'= -C$_2$H$_5$;

81b: R= -C$_6$H$_5$, X=O, R'= -C$_2$H$_5$; **82c**: R= -C$_6$H$_5$, X=O, R'= ◁; **82a**: R'= -C$_2$H$_5$; **82b**: R'= ◁

Scheme (18). Microwave mediated synthesis of fluoroquinolone-azole-thiomorpholine hybrids.

The replacement of thiomorpholine ring by 4-(2-methoxyphenyl)piperazine moiety and the character of linker between fluoroquinolone skeleton and azole unit has not caused a change in antibacterial activity [2].

In another study reported by Mentese *et al*, the 1,2,4-triazole and norfloxacin units have been merged by a series of condensation and cyclisation reactions on fluoroquinolone skeleton with the assistance of microwave (Scheme **19**) [101].

Hybridization with Oxadiazole

1,3,4-Oxadiazole ring has been defined as another heterocyclic unit having a privileged structure in the medicinal chemistry [29, 101] and their anticancer and antibacterial activities have been emphasized by numerous studies [6, 29, 113 - 117].

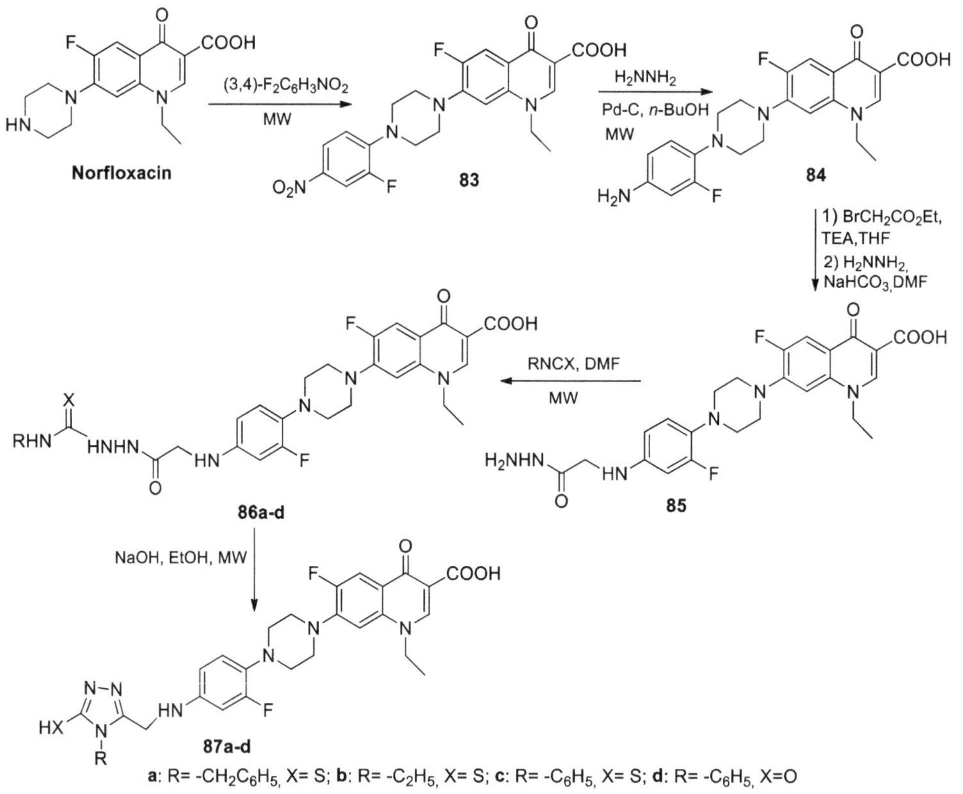

Scheme (19). Cyclisation of carbo(thio)amides to triazole ring at the C-7 position of norfloxacin.

The cyclization of the compounds having thiosemicarbazide structure has shown to be an excellent strategy for the synthesis of 1,2,4-triazole, 1,3,4-thiadiazole and 1,3,4-oxadiazole and derivatives [91]. Alternatively, the cyclodehydration of arylcarboxylic acids with semicarbazide in the presence of phosphorous oxychloride produces the corresponding 2-amino-5-aryl-1,3,4-oxadiazoles [118]. Kumar *et al.* substituted the 1,3,4-oxadiazole unit to the *N*-4 position of piperazin ring in fluoroquinolones *via* the formation of diazonium salt of 2-amino-1,3-4-oxadiazole compound with the purpose to generate new antibacterial compounds [118]. The diazotation of 2-amino-5-aryl-1,3,4-oxadiazoles by nitrous acid has led to the formation of the corresponding diazonium chloride salt. Further, the addition of finely divided copper powder into it resulted in the formation of 2-chloro-5-aryl-1,3,4-oxadiazoles, which have then been subjected to integrate with fluoroquinolones (Scheme **20**).

88a-92a: Ar= -C$_6$H$_5$, **88b-92b**: Ar= 3-NO$_2$C$_6$H$_4$; **88c-92c**: Ar= 4-NO$_2$C$_6$H$_4$; **88d-92d**: Ar= 3,4-diNO$_2$C$_6$H$_3$;

88e-92e: Ar= 3-OMeC$_6$H$_4$; **88f-92f**: Ar= 3,4-diOMeC$_6$H$_3$; **88g-92g**: Ar= 2-ClC$_6$H$_4$; **92a**: R$_1$= R$_3$= H,

R$_2$= -C$_2$H$_5$; **92b**: R$_1$=R$_3$= H, R$_2$= cyclopropyl; **92c**: R$_1$= -CH$_3$, R$_2$= cyclopropyl, R$_3$= OCH$_3$

Scheme (20). Synthetic pathway for the preparation of oxadiazole-quinolone hybrids.

Another fluoroquinolone-oxadiazole hybrid (**94**) in which the two pharmacophores were connected *via* a fluorophenylene linker, has been reported by Mentese *et al.* [101]. Then this compound has been converted into the Mannich base (**95**) with a reaction prompted by microwave (Scheme **21**).

Scheme (21). Microwave prompted synthesis of oxadiazole-fluoroquinolone hybrids.

Hybridization with Oxazolidinones and Thiazolidinones

Oxazolidinones, which act through a unique mechanism of action containing bacterial protein synthesis inhibition, constitute a new class of completely synthetic antibacterial agents (Fig. **12**).

Fig. (12). Linezolid and Eperezolid.

Zyvox (linezolid) that is the first example of oxazolidinone class antibacterial drugs, has been widely used for the treatment of the infectious diseases caused by Gram-positive pathogens. Although first generation oxazolidinones have generally exhibited activity on Gram positive bacteria, the enhanced spectrum and improved potency of new second generation oxazolidinones possessing activity on Gram negative bacteria could broaden the efficacy of this class of antibacterials in the clinical settings for the treatment of infectious diseases [119, 120]. It has been accepted that the oxazolidinone–quinolone hybrids exhibit a balanced dual mode of action containing the inhibition of DNA gyrase and topoisomerase IV, and protein synthesis inhibition.

Based on this, the integration of these two pharmacophores has attracted a great deal by synthetic and medicinal chemists as a general strategy leading to the preparation of new antibacterials with expanded activity spectrum [121]. New oxazolidinone-fluoroquinolone hybrids have been designed by Hubschwerlen *et al.* as potent antibacterial compounds (Scheme **22**) [120, 121].

Hubschwerlen *et al.*'s studies have revealed that hybrids containing the oxazolidinone–quinolone integration (**96a-n**) simultaneously act on two different cellular functions, DNA replication (DNA gyrase and topoisomerase IV) and protein synthesis. It is important to note that the hybrids which differ only in the nature of the linker, exhibit evidently different antibacterial potency [120]. Among the hybrid (**96a-n**), the compound with a piperazinyl spacer (**96a**) has displayed strong oxazolidinone characteristics besides some quinolone-like properties including good activity on ciprofloxacin-resistant strains. Moreover, **96a** has behaved as a stronger protein synthesis inhibitor than linezolid. On the other hand, the hybrid with 3-amino-pyrrolidinyl linker (**96b**) has shown quinolone-like behavior, such as high susceptibility against linezolid-resistant S. aureus and Gram negative strains, and resistance to ciprofloxacin, while the hybrids connecting *via* 3-hydroxy-azetidinyl or a 3-hydroxy-pyrrolidinyl linker (**96c-e**) has exhibited a balanced dual action mode with good activity towards both ciprofloxacin- and linezolid-resistant strains. Also, the spacers 4-hydrox--piperidinyl or a 4-hydroxy-azepanyl (**96h, i**) have afforded the balanced mode of activity on both linezolid- and ciprofloxacin-resistant strains. Further studies on the effect of the length of the spacer have shown that this factor has no significant impact on the activity. On the other hand, the *R* configuration at position 3 on the pyrrolidine has slightly enhanced the quinolone character, while the *S* configuration has increased the oxazolidinone character [120, 121].

Scheme (22). Preparation of new oxazolidinone-fluoroquinolone hybrid compounds.

Scheme (23). Design of oxazolidinone-ciprofloxacin hybrids.

The design of the binding positions of oxazolidinone and quinolone pharmacophores has critical importance for antibacterial activity [119]. The type of **104** hybrids containing *N*-1 connected oxazolidinone unit has been found to be inactive, while other type hybrids connecting at the C-7 position of fluoroquinolone core *via* a piperazinyl group has exhibited quite high activity towards Gram-positive strains [119]. Moreover, hybrids **98, 101** and **102** were found to be 2–4 times more active than linezolid against two gram-negative respiratory tract pathogens, Haemophilus influenzae and Moraxella catarrhalis. On the other hand, these compounds have not displayed activity as the parent quinolones. In the Gordeev's study, the (*S*)-enanatiomer **102** have been found to be somewhat more active than their rasemic form **101**, as expected from SAR studies. This parallels in the ancestor parent quinolone series, wherein the 3-(*S*)-enantiomer of levofloxacin has higher activity than ofloxacin that can be regarded as racemic analog of it (Scheme **23**, Scheme **24**) [119].

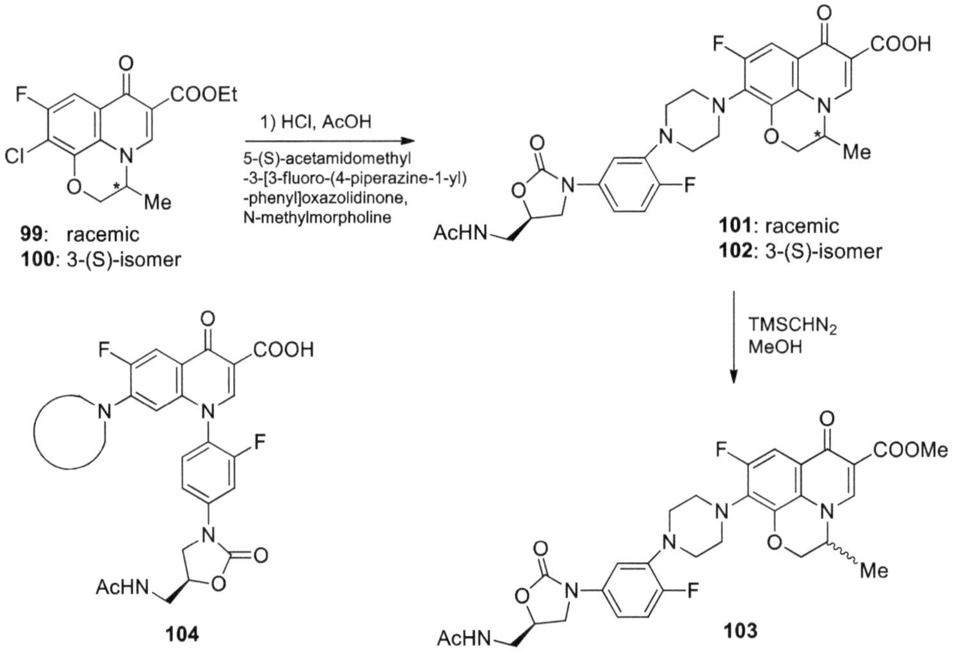

Scheme (24). Design of oxazolidinone-levofloxacin hybrids.

Among the quinolone-oxazolidinone hybrids, the most promising compounds MCB3681 and MCB3837 have exhibited excellent activity for the treatment of several infections caused by various microorganisms including drug-resistant and multidrug resistant pathogens (Fig. **13**) [50].

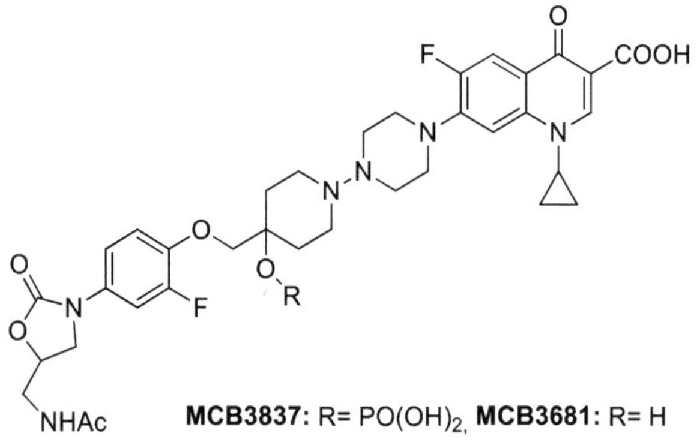

Fig. (13). MCB3681 and MCB3837.

Cadazolid (ACT-179811; Actelion Pharmaceuticals) has been developed as an oxazolidinone- quinolone hybrid that recently started a phase-III trial for the treatment of Clostridium difficile infections (Fig. **14**).

Cadazolid

Fig. (14). Cadazolid.

Thiazolidinone core that constitutes another privileged structure in the class of *N* and *S* containing heterocycles with antibacterial activity, has been attracting extensive attention in the field of combinatorial synthesis [2, 99]. Thiazolidinones have also been reported to be responsible for nematocidal, fungicidal, antibacterial and antiviral activities due to the presence of the *N-C-S* linkage [100].

The norfloxacin-thiazole and norfloxacin-thiazolidinone hybrids with potent antibacterial activity have been reported by Mentese *et al.* (Scheme **25**) [101].

Scheme (25). Cyclisation of carbo(thio)amides to thiazole ring at C-7 position of norfloxacin.

Hybridization with Thiazole

2-Aminothiazolyl group as an important structural fragment has expeditiously found application in the design and synthesis of drugs. Numerous 2-aminothiazole drugs have been successfully developed and widely used for clinical treatment of various diseases, such as antibacterial, anticancer, antiparasitic, anti-inflammatory. Especially, lots of 2-aminothiazole derivatives like Ceftazidime, Cefuzonam, Carumonam and Aztreonam have played an important role in fighting against infective diseases as antimicrobial agents (Fig. **15**).

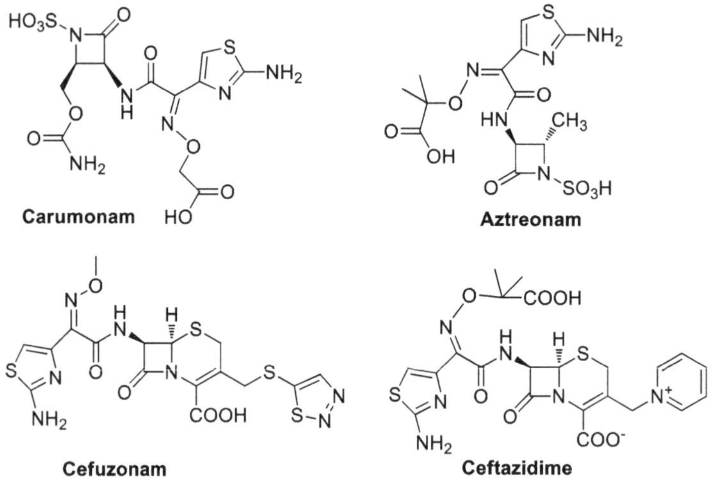

Fig. (15). Ceftazidime, Cefuzonam, Carumonam and Aztreonam.

The clinical success of 2-aminothiazole drugs has led to emerge an array of new bioactive molecules. In order to improve biological activities, the introduction of 2-aminothiazolyl group to the existing drugs has become one of the most common approaches. In this connection, Cheng *et al.* demonstrated that the introduction of 2-aminothiazole nucleus to fluoroquinolone skeleton led to discover promising drug candidates (Scheme **26**) [67].

Further investigations on thiazole-quinolone hybrids have demonstrated that the incorporation of aminothiazole unit into C-3 position in quinolones causes an increase in the antibacterial activity, while the introduction of thiazole nucleus into C-7 position leads to diminish of the antibacterial activity [50]. On the other hand, the fusion of thiazol-4(5*H*)-one unit into 2,3-position of quinolone core has afforded new hybrids with antibacterial activity (Fig. **16**) [50].

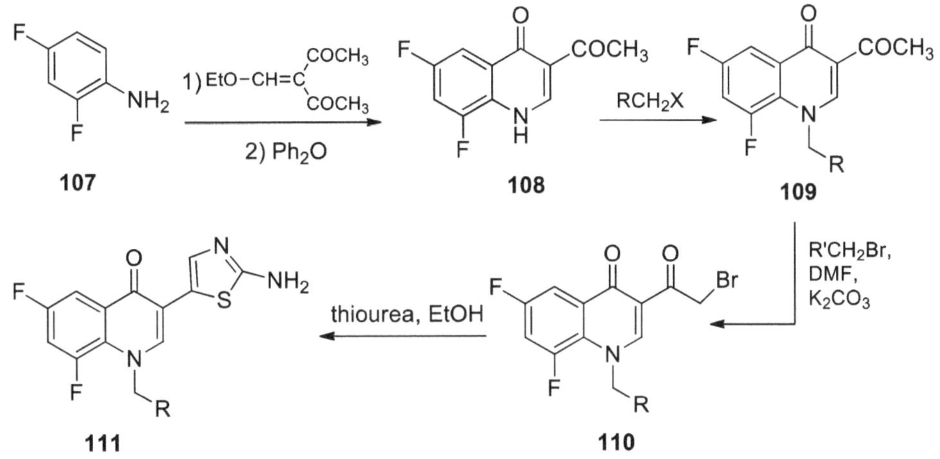

Scheme (26). Synthesis of C3-aminothiazole substituted quinolones.

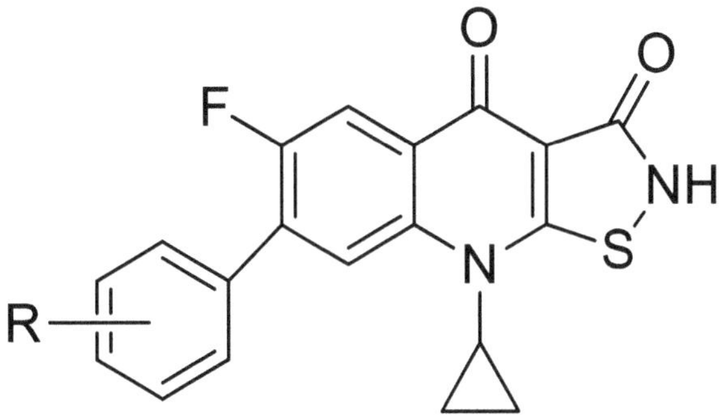

Fig. (16). 7-Aryl-isothiazolo[5,4-b]quinoline-3,4(2*H*,9*H*)-dione.

The SAR studies on isothiazolo[5,4-b]quinolone derivatives **112** have demonstrated that the position of substituent on the phenyl group should be OH or NH_2. The modification of these substituents results in the reduction of activity [122].

Novel fluoroquinolone-benzothiazole annulates having promising antibacterial activity has been reported by Sharma *et al.* (Scheme **27**) [123].

R₁= -OCH₃, -CH₃, -NO₂, -F, -Br; R₂= cyclopropyl, ethyl; R₃= -H, -OCH₃; R₄= -H, -CH₃

Scheme (27). Novel fluoroquinolone-benzothiazole annulates.

Hybridization with Benzimidazole

Benzimidazole ring has drawn much attention by synthetic organic and medicinal chemists due to its prevalent biological activities, especially antimicrobial aspect, and benzimidazole derivatives have been revealed to be specific topoisomerase inhibitors [124 - 128].

Zhang and coworkers have designed new DNA-binding antimicrobial agents consisting of two pharmacophores namely benzimidazole and fluoroquinolone, which are connected to each other *via* a flexible linker (Scheme **28**) [128].

Hybridization with Pyrazole

Quinolone-pyrazol hybrids have displayed significant *in vitro* antibacterial activities towards Gram-positive organisms, while they have shown weak inhibition on Gram-negative strains (Fig. **17**) [25].

Scheme (28). Synthesis of fluoroquinolone-benzimidazole conjugates.

R_1= -H, -CH$_3$; R_2= cyclopropyl, 2,4-difluorophenyl;ethyl,-CH$_2$CH$_2$F;

X= N, C, COMe, COCHF$_2$

Fig. (17). 7-(3-Amino-2-alkylpyrazolo[3,4-c]pyrrol-1-yl)-fluoroquinolones.

Oxime Functionalized 4-Quinolones

A number of oxime-functionalized azetidines, pyrrolidines or piperidines have been introduced into the C-7 position of quinolones, and some of them have been found to possess excellent *in vitro* and *in vivo* antibacterial activities with improved pharmacokinetic and safety profiles [53, 129 - 131]. Recently, oxime-functionalized pyrrolidines as novel C-7 substituents have attracted great attention and led to the discovery of some new fluoronaphthyridones, such as gemifloxacin, and zabofloxacin. In this connection, IMB-070593, IMB-070595 and 85 have been reported by Lv and coworkers as potent antibacterial oxime functionalized quinolone antibiotics. These compounds have shown excellent activity and pharmacokinetic profiles which emphasize the importance of the oxime group with respect to biological activity (Fig. **18**) [53].

Gemifloxacin

Zabofloxacin

DZH

IMB-070593 (X= C-OMe), IMB-070595 (X=N)

122: Y: F or H

Fig. (18). Oxime-functionalized pyrrolidines.

Chalcon Functionalized Quinolones

Whether they are natural or synthetic, chalcones have attracted interest due to their biological properties. It is reported that chalcones have the ability to inhibit cancer cell proliferation, induce apoptosis and display exceptional impact on skin carcinogenesis [132]. It is well documented that the introduction of a substituent to *N*-4-piperazinyl core of quinolones causes to alter physicochemical properties of quinolones affecting cytotoxicity and selectivity against topoisomerase I and II [132]. Abdel-Azez *et al.* have reported the synthesis of new C-7-piperazin-1-fluoroquinolone-chalcone hybrid molecules with anticancer and/or topoisomerase I and II inhibition activity (Scheme **29**) [132].

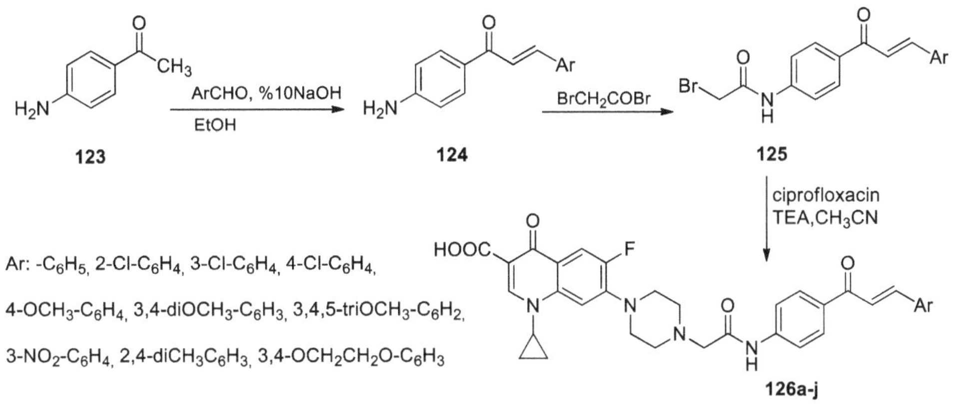
ArCHO, %10NaOH
EtOH
BrCH2COBr
ciprofloxacin
TEA,CH3CN
123
124
125
126a-j
Ar: -C6H5, 2-Cl-C6H4, 3-Cl-C6H4, 4-Cl-C6H4,
4-OCH3-C6H4, 3,4-diOCH3-C6H3, 3,4,5-triOCH3-C6H2,
3-NO2-C6H4, 2,4-diCH3C6H3, 3,4-OCH2CH2O-C6H3

Scheme (29). Synthesis of C-7-piperazinyl-fluoroquinolone-chalcone hybrid molecules.

Hybridization with Metronidazole

The imidazole derivatives containing a nitro group have been found to possess antimicrobial activity, and many of them such as benznidazole, secnidazole, metronidazole, nimorazole and ornidazole (Fig. **19**) have been extensively used for the treatment of infections caused by anaerobic pathogens.

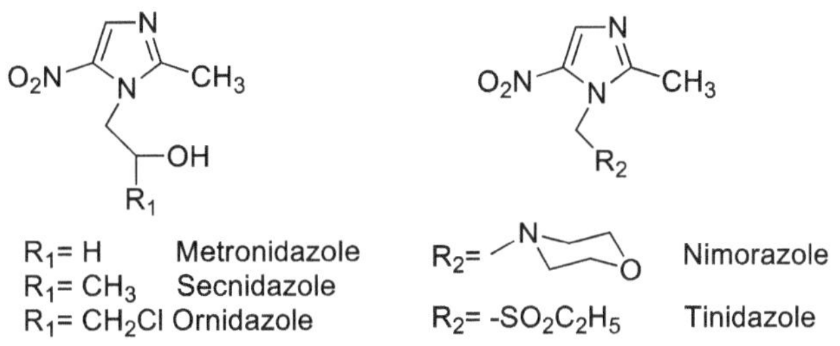
R1= H Metronidazole
R1= CH3 Secnidazole
R1= CH2Cl Ornidazole
R2= Nimorazole
R2= -SO2C2H5 Tinidazole

Fig. (19). Drugs containing imidazole core.

The nitro group, that constitutes an important unit leading to enhance lipophilicity, has been accepted as a favorable fragment for tissue penetration. Moreover, the metabolic activation of the nitro group can improve the biological activity of target compounds. These distinguished achievements have served many researchers to focus on the design and synthesis of new metronidazole derivatives having potential clinical applications. A number of studies on mechanism have disclosed that the reactive metabolic intermediates formed in bacteria by the reduction of nitro group in nitroimidazoles can covalently bind with DNA and trigger the adverse effects. Thus, the steric protection of nitro group in metronidazole has been proved to be an efficient strategy to improve the metabolism and physicochemical properties of new compounds [133, 134]. Zhang *et al.* have been designed new quinolones, which are combined with metronidazole and its 4-cyclic amino derivatives at the C-7 position in order to produce a new hybrid scaffold of antibacterial compounds (Scheme **30**) [133].

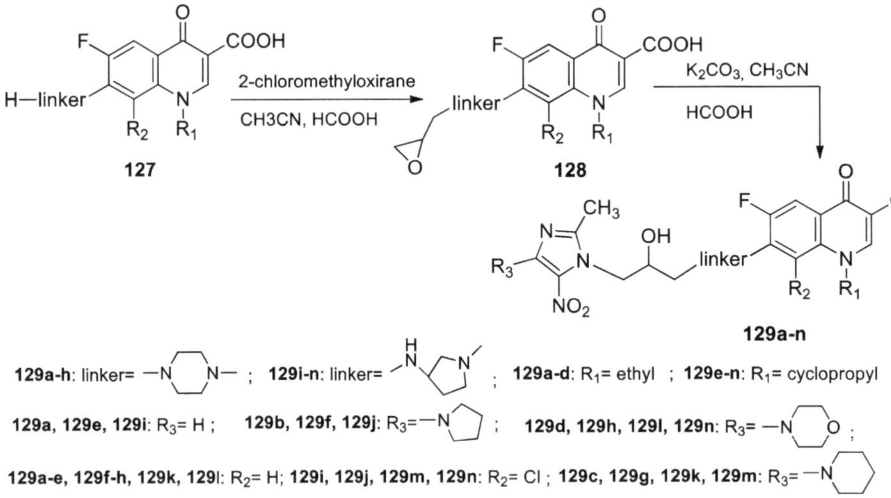

Scheme (30). Hybridization of quinolones with metronidazole at C-7.

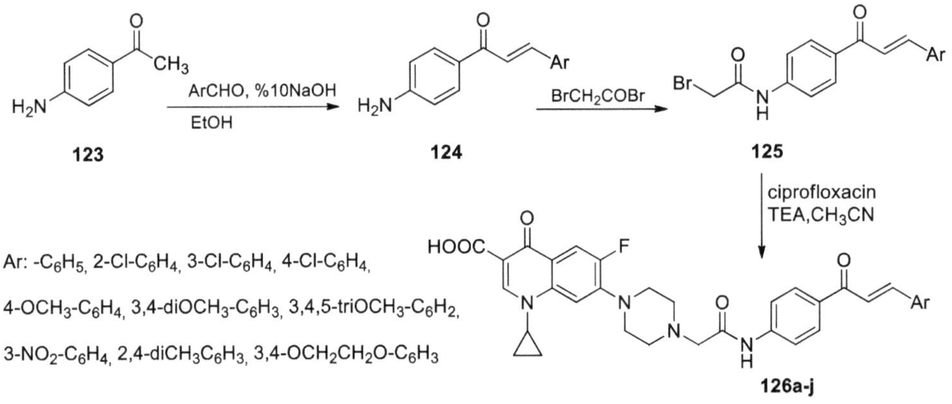

Scheme (29). Synthesis of C-7-piperazinyl-fluoroquinolone-chalcone hybrid molecules.

Scheme (31). Hybridization of quinolones with metronidazole at N-1.

Cui *et al.* have introduced the nitroimidazole moieties into *N*-1 position of quinolones to generate a new series of metronidazole-quinolone hybrids, because the presence of nitroimidazole group at *N*-1 position has been expected to enforce non-covalent interactions with DNA base or topo-DNA complex to get over the drug resistance and extend the antibacterial spectrum (Scheme **31**) [134].

Hybridization with Isatin

Another naturally occurring scaffold, isatin has attractedwidespread attention because of its various pharmaceutical properties including anti-microbial, anti-viral, anti-cancer, anti-HIV, anti-oxidant, and central nervous system depressant [135 - 140]. The compounds containing isatin structural motif have also been reported as potent DNA gyrase inhibitors. Its stability and structural viability have allowed it to integrate with different pharmacophore groups leading to access to new bioactive hybrid compounds.

Fig. (20). Fluoroquinolones containing an isatin unit.

Therefore, isatin has emerged as a rational scaffold which is a suitable new bioactive molecules. In this connection, Xu *et al.* have obtained new anti-tuberculosis agents integrating isatin unit to the C-7 position of some fluoroquinolones and C-3 position of nalidixic acid (Fig. **20**) [141].

Vandekerckhove *et al.* have reported new quinolone-indole hybrids with antiplasmodial activity (Scheme **32**) [32].

Pyrazine and Isoxacine Hybrids

The integration of quinolones with pyrazine has been shown to cause the emergence of considerable antibacterial potential, and the SAR studies indicated that the activity has a positive correlation with the length of the linker. The optimum length of n is 2 or 3, while in the case of longer linkers, the activity losses completely (Fig. **21**) [50].

Pyrazinamide, that is one of the first-line antituberculosis drugs, constitutes an important part for the treatment of tuberculosis. Pyrazinamide derivatives have been found to be active against even drug-resistant tuberculosis. In addition, amino acids have gained importance as drug candidates due to their permeability into mammalian tissue. The pro-drugs consisting of amino acid and quinolone units are more lipophilic than the parent compounds and therefore display improved *in vitro* antibacterial activities [50]. New fluoroquinolone-pyrazine conjugates connected to each other *via* amino acid linkers have been synthesized by Panda and his group as antimicrobial agents (Scheme **33**) [142].

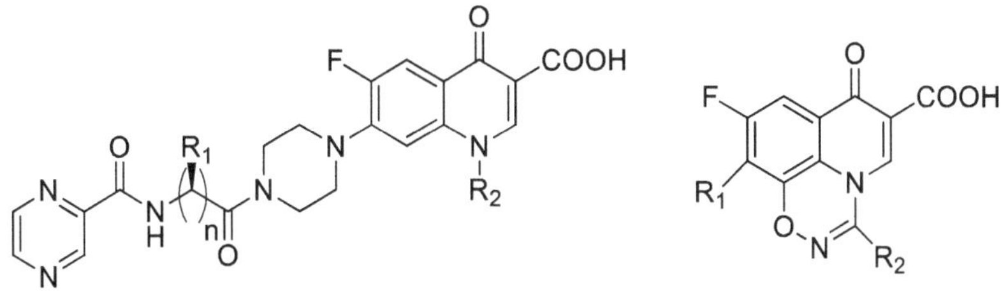

147

n= 1,2,3; R_1= H, Me; R_2= cyclopropyl, ethyl

148

R_1= H, Me

Fig. (21). Pyrazine and isoxacine Hybrids.

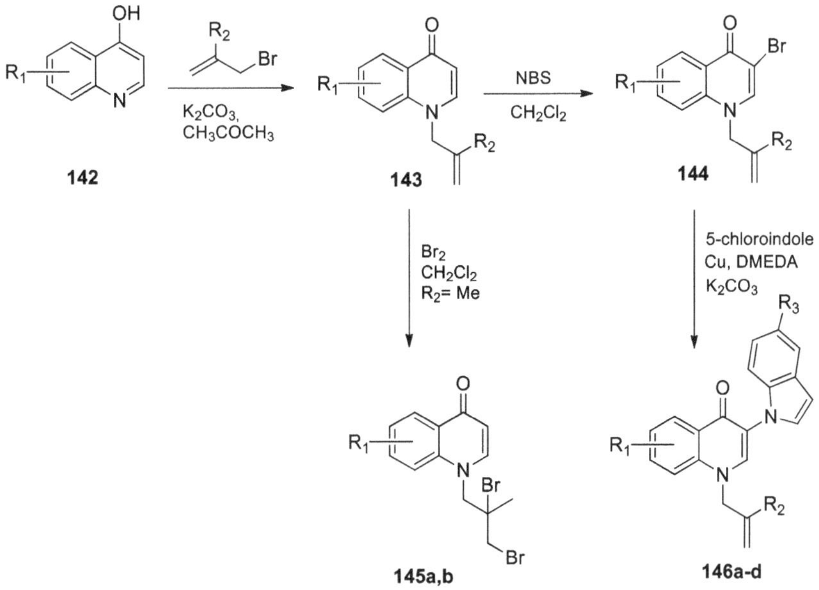

142

143

144

145a,b

146a-d

R_1= H, 7-Cl, 3-CO_2Et, 8-F; R_2= H, CH_3; R_3= H, Cl

Scheme (32). Synthesis of quinolones containing indole unitat C-3 position.

a: R₁=H; n= 1, R₂= ethyl; b= R₁= Me, n= 1, R₂= ethyl; c= R₁= H, n= 2, R₂= ethyl; d: R₁= H, n= 3, R₂= ethyl; e: R₁=H; n= 1, R₂= cyclopropyl; f: R₁=CH₃; n= 1, R₂= cyclopropyl; g: R₁= H, n= 2, R₂= cyclopropyl; h: R₁= H, n= 3, R₂= cyclopropyl

Scheme (33). New fluoroquinolone-pyrazine conjugates with amino acid linkers.

Pyrimidine Hybrids

Pyrimidines have been known as medicinally important heterocyclic compounds with a wide range of biological activities [143 - 146]. The SAR studies on the ciprofloxacin-pyrimidine hybrids in which these two units have been connected through an NH, NHCH₂ or 1,2,3-triazole linkers have shown that the character of linker strongly affects the activity. The efficiency has diminished in this order; 1,2,3-triazole> NH > NHCH2. The activity has also been affected by a linked pattern of triazole, and the directly linked pattern at C-7 has been found to be more affirmative than those linked to piperazinyl subunit and methylene (Fig. **22**) [145].

Fig. (22). Structures of linker-connected pyrimidine-quinolone hybrids.

It is noteworthy that the hybrids exhibit dual-acting mechanisms: inhibiting the targets of both trimethoprim (dihydrofolate reductase) and quinolone (DNA gyrase and topoisomerase IV).

The lead hybrid BP-4Q-002 has been found to be comparable to or more potent than trimethoprim, ciprofloxacin alone or 1:1 a mixture of ciprofloxacin and trimethoprim against drug-sensitive and CPFX-resistant bacteria [146].

Hybridization with Flavonoids

Flavonoids have been widely found in natural sources including fruits, vegetables, seeds and herbs and they belong to a phenolic class of organic compounds. *In vivo* studies have proven that flavonoids have a protective effect in the prevention of many diseases, such as cancer, cardiovascular disease and neurodegenerative disorders [81, 147]. Due to this reason, flavonoids have become the center of much of the medicinal chemistry studies devoted to the discovery of new bioactive compounds with improved properties [38]. Apigenin homodimers (**160**) have been reported as modulators of the efflux pump P-glycoprotein activity in human cancer [147 - 154]. Tetramethylscutellarein (**161**) and sarothrin (**162**) have been reported to possess potent inhibitory activity of bacterial NorA efflux pump [38]. Several other flavonoids including quercetin (**163**), chrysin (**164**) and genistein (**165**) have also been defined as efficient inhibitors of the multidrug transporters from the MRP Family (Fig. **23**) [38]. Inspired by the inhibition capacity of some flavonoids towards efflux pumps, and based on the concept of "molecular hybridization", Xiao *et al.* synthesized some flavonoid-floxacin conjugates in order to address the serious issues (Scheme **34**) [38].

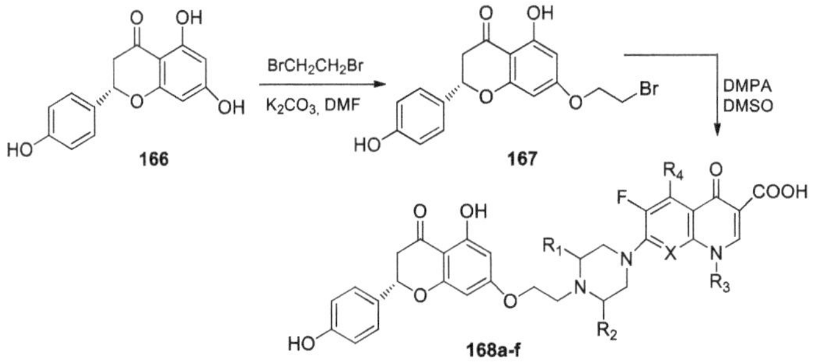

a: $R_1=R_2=R_4=$ H, $R_3=$ ethyl, X= CH; **b**: $R_1=R_2=R_4=$ H, $R_3=$ 4-fluorophenyl, X= CH;
c: $R_1=$ methyl, $R_2=R_4=$ H, $R_3=$ ethyl, X= CF, **d**: $R_1=R_2=R_4=$ H, $R_3=$ cyclopropyl, X= CH;
e: $R_1=R_2=R_4=$ H, $R_3=$ ethyl, X= N, **f**: $R_1=R_2=$ methyl, $R_4=$ NH_2, $R_3=$ cyclopropyl, X= CF

Scheme (34). Synthetic pathway for the synthesis of flavonoid-fluoroquinolone hybrids.

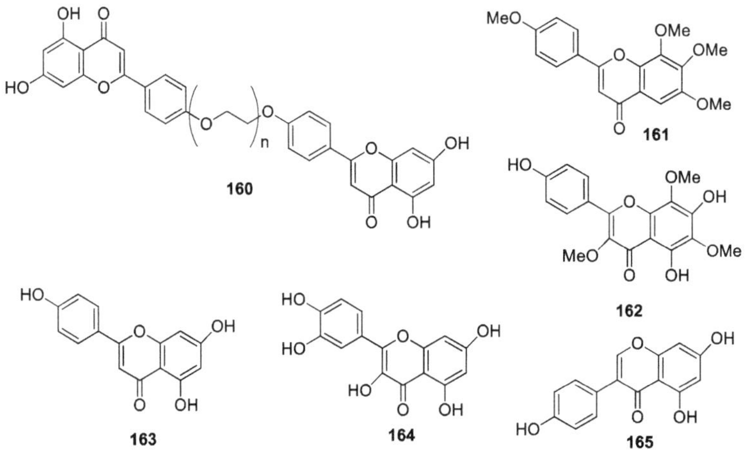

Fig. (23). Some known flavonoids with biological activity.

Quinolone-Quinolone Conjugation

A series of quinolone–fluoroquinolone conjugates connected *via* an amino acid linker have also been synthesized by Panda *et al.* (Schemes **35, 36**) [27].

Scheme (35). Fluoroquinolones with aminoacid function.

Scheme (36). Synthesis of linker connected quinolone-quinolone conjugates.

It is expected from such a conjugation to decrease the degradation of antibiotics and hence increase the concentration at the target site. The initial dosage may be lowered if more antibiotic molecules reach the target. This means that the new analogs may have the capacity to maintain or improve activity and also diminish adverse effects [27]. The amino acid-linker in these derivatives has been used to increase their permeability into mammalian tissue. Because prodrugs formed from quinolone acids and amino acid esters have been accepted to be more lipophilic than the parent drugs and can display improved *in vivo* antibacterial activity with prominent therapeutic effects. Because cell penetration of antibiotics is accepted to be important in the elimination of infections caused by intracellular pathogens [27].

Kerns *et al.* reported the synthesis of new linker-connected quinolone dimers with antibacterial potency towards fluoroquinolone resistant strains (Fig. **24**) [155]. According to Kern's studies, the fluoroquinolone monomer itself and the linker connected to the dimer has displayed the inhibition activity. The linker that leads to the most active dimer of a certain fluoroquinolone does not have to be the optimal linker for other symmetric dimers of different fluoroquinolones [155]. In order to improve the bacterial permeability, Ross *et al.* designed the peptide-linked and polyethylene glycol (PEG)-linked dimers of ciprofloxacin with appropriate physicochemical properties (Fig. **24**) [156].

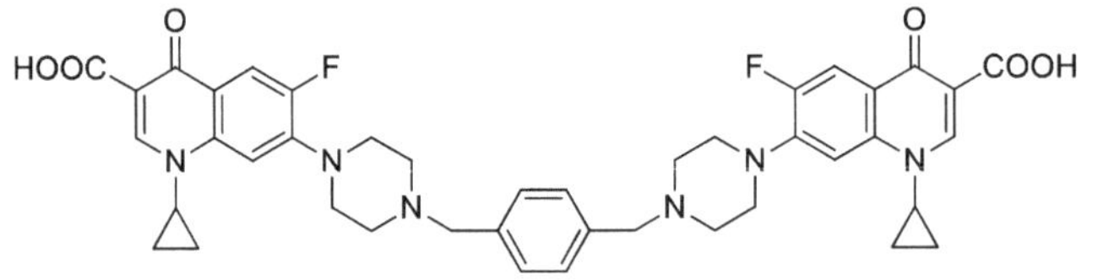

176

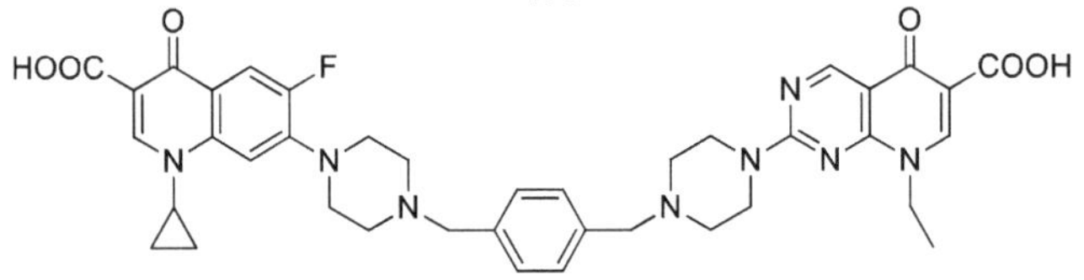

177

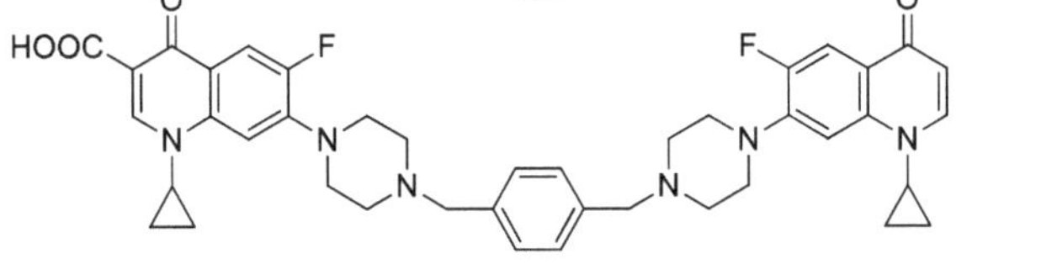

178

Fig. (24). Linker connected quinolone dimers.

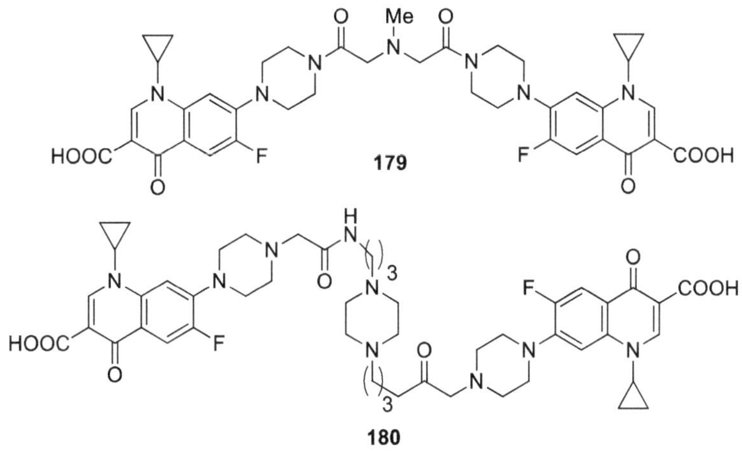

Fig. (25). Peptide-linked and PEG-linked dimers of ciprofloxacin.

Khan *et al.* prepared novel fluoroquinolone dimers which are connected at *N*-1 position *via* an ethylene linker (Fig. **26**) [157].

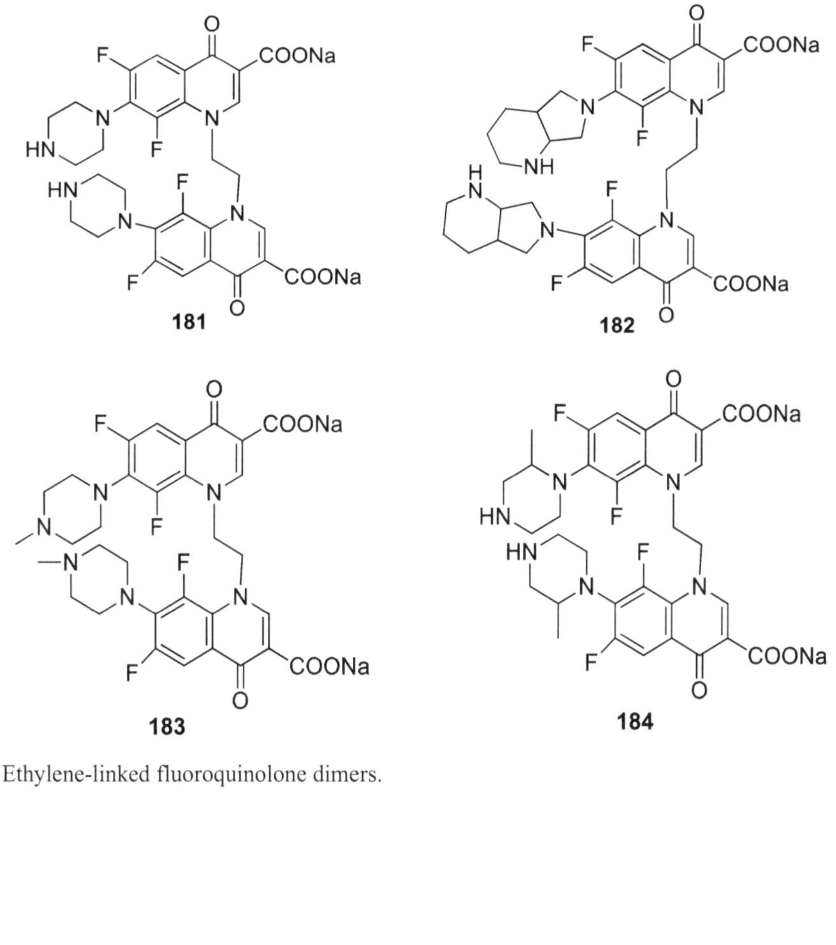

Fig. (26). Ethylene-linked fluoroquinolone dimers.

Former studies revealed that it is not necessary to remain the carboxylic acid function at the position-3 in the quinolone skeleton for biological activity, and it can be replaced by heterocyclic rings, such as 1,2,4-triazole and triazolothiadiazole core. Hu and coworkers have prepared new fluoroquinolone dimers with antitumor activity from the integration of monomers *via* a 1,2,4-triazolo[2,1-b] [1, 3, 4]thiadiazole linker at the C-3 position (Fig. **27**) [158].

Fig. (27). Triazolothiadiazole connected fluoroquinolone dimers.

Amidation and Substitution By Homoaromatic Ring at C-3

The synthesis of powerful topoisomerase I inhibitors has been achieved *via* homocyclic ring substitution at C-3 (Scheme **37**) [55].

Scheme (37). Synthesis of fluoroquinolones containing homoaromatic ring at C-3.

Besides their antibacterial activity, quinolones have reactive oxygen species (ROS) scavenging ability [159, 160]. Quinolone scaffold has also been efficiently used for the design of dual acetyl colin esterase (AChE) inhibitör [161, 162]. As it has been well known, Alzheimer's disease (AD) constitutes the primary reason for dementia impressing roughly 10% of the population over the age of 65 years old, and its ration increases exponentially with age. Acetylcholinesterase has constituted the basic target of the therapeutic approach to treat cognitive loss associated with AD. Because, AD has been accepted to be associated with deterioration in cholinergic transmission. In order to design AChE inhibitors with improved properties as a therapeutic agent that will delay the progression of neurodegeneration, scientists have devoted much effort to the development of new multipotent compounds [162 - 164]. Moreover, multitargeted drugs having AChE inhibition and antioxidant activity combination have emerged as a another key target to delay AD's progression, because the oxidative stress plays a central role in AD pathogenesis [165]. Based on these, Pudlo *et al.* have carried out the synthesis of quinolone derivatives with AChE inhibition and antioxidant activity by the amidation of acid function at C-3 in 2-quinolone scaffold (Fig. **28**) [162].

$$R_1 = H, Me, Bn; \quad R_2 = OH, Me, Ph, NH_2; \quad n = 0, 1$$

Fig. (28). Quinolones with anti-AChE and antioxidant activity.

Hybridization with β-Lactams

In this class, C-3 linked hybrids have been shown to be more active than C-7 linked derivatives. In particular, the hybrid Ro-23-9424 has exhibited promising activity on both drug-sensitive and drug resistant bacteria including cephalosporin- and FQs-resistant pathogens, probably through a new mechanism (Fig. **29**) [50].

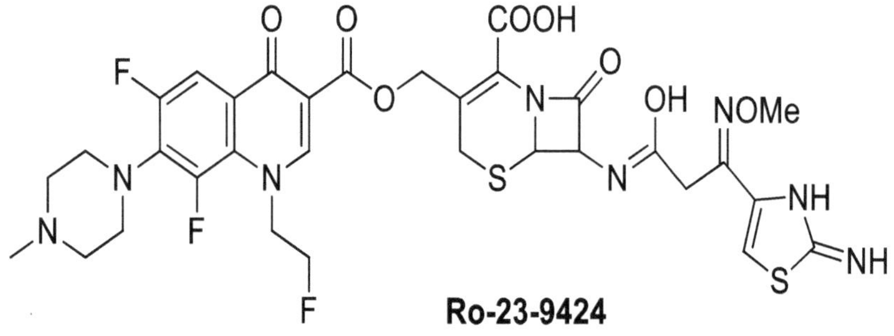

Fig. (29). The stucture of Ro-23-9424.

Hybridization with Macrocycles

In order to discover better therapeutic agents on persistent infectious diseases, a series of rifamycin based bi-functional compounds have been designed, synthesized and evaluated for their activity. In this connection, quinolone-macrocycles conjugates with dual-mode of action have been prepared with the goal to maintain the potent activity of rifamycins against persistent pathogens and at the same time minimize the development of rifamycin resistance (Scheme **38**) [50].

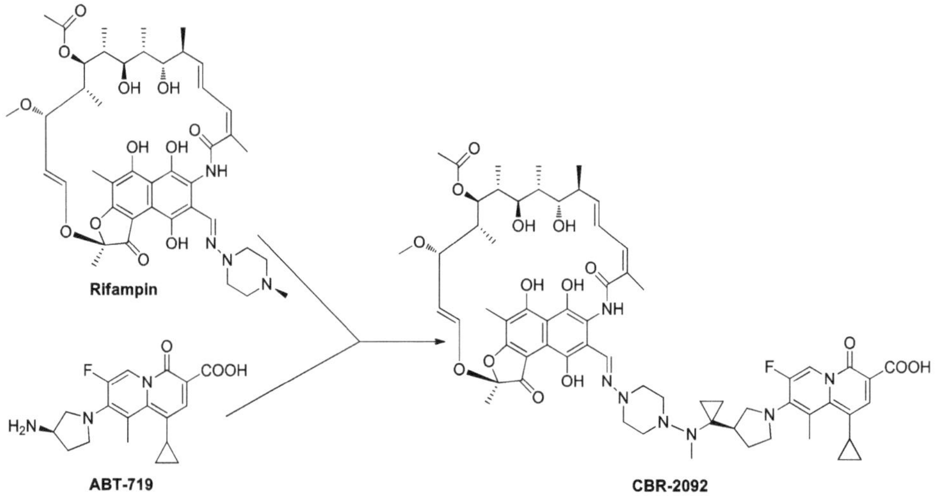

Scheme (38). Rifampin-quinolone hybridization.

Rifamycin class antibiotics have displayed potent activity on an array of Gram (+) strain, including mycobacteria as well, and have been used for the treatment of the infections caused by *Mycobacterium tuberculosis*. The *in vitro* biofilm assays and animal models of biofilm-associated infections have demonstrated that rifamycins, alone or in combination are active towards susceptible bacteria disseminated in the biofilm state. This has served the scientists to admit rifampin-containing regimens as standard therapies for the treatment of biofilm-associated infections. The rifamycin-quinolone hybrid CBR-2092, that has been obtained by the combination of the rifamycin with a quinolone pharmacophore (ABT-719) and is under phase II clinical trial, exhibited high activity on FQs-resistant and multi-drug resistant pathogenic microorganisms probably *via* a different mechanism of action from FQs. This is important to overcome drug resistance [166 - 169].

Aminoglycoside antibiotics, which target bacterial protein synthesis machinery, constitute another clinically important class of drugs suffering largely due to resistance [170]. In order to deal with the problem of drug resistance caused by enzymatic modification, some analogs of aminoglycoside antibiotics have been synthesized by structural modification of the existing aminoglycosides. Previous efforts in this direction have resulted in several semisynthetic drugs including amikacin, dibekacin and arbekacin (Fig. **30**).

amikacin dibekacin arbekacin

Fig. (30). Some known aminoglycoside antibiotics.

However, after their introduction to the clinic, new resistance to these drugs has emerged. As one of the efforts aiming to overcome this undesirable case, a combination of ciprofloxacin and Neomicin B (NeoB) has been designed where the two pharmacophoric units are linked *via* various spacers: spacer X comes from ciprofloxacin nucleus and spacer Y from NeoB nucleus. These hybrids have exhibited the dual mode of action by inhibiting both targets: bacterial protein synthesis and topoisomerase/ gyrase. (Scheme **39**) [170].

Y= -CH₂-, -*p*-C₆H₄; X= -(CH₂)₂-, -(CH₂)₃-, -(CH₂)₄-, -(CH₂)₅-, -(CH₂)₆-, -CH₂CHOHCH₂-, -(CH₂)₂-O-(CH₂)₂-, -CH₂-*m*C₆H₄-CH₂-,-CH₂-*p*C₆H₄-CH₂-

Scheme (39). Neomicin B-quinolone hybridization.

CONCLUDING REMARKS

Rapidly growing drug resistance towards commonly used antibiotics has made the development of new drugs a necessity, and this has stimulated the scientists to design and discover new agents consisting of two or more pharmacophore groups or drug fragments each with different modes of action. This multiple target strategy has resulted in the discovery of hundreds of hybrid molecules also containing drug dimerization or drug-drug conjugation. Some of the hybrid compounds exhibited an entirely new mechanism, while others exhibited a dual mode of action.

CONSENT FOR PUBLICATION

Not applicable.

CONFLICT OF INTEREST

The authors confirm that this chapter content has no conflict of interest.

ACKNOWLEDGEMENTS

Decleared none.

REFERENCES

[1]   Basoglu, S.; Demirbas, A.; Ulker, S.; Alpay-Karaoglu, S.; Demirbas, N. Design, synthesis and biological activities of some 7-aminocephalosporanic acid derivatives. *Eur. J. Med. Chem.,* **2013**, *69*, 622-631.
        [http://dx.doi.org/10.1016/j.ejmech.2013.07.040] [PMID: 24095755]

[2]   Mentese, M.Y.; Bayrak, H.; Uygun, Y.; Mermer, A.; Ulker, S.; Karaoglu, S.A.; Demirbas, N. Microwave assisted synthesis of some hybrid molecules derived from norfloxacin and investigation of their biological activities. *Eur. J. Med. Chem.,* **2013**, *67*, 230-242.
        [http://dx.doi.org/10.1016/j.ejmech.2013.06.045] [PMID: 23871903]

[3]   Demirci, S.; Demirbas, A.; Ulker, S.; Alpay-Karaoglu, S.; Demirbas, N. Synthesis of some heterofunctionalized penicillanic acid derivatives and investigation of their biological activities. *Arch. Pharm. (Weinheim),* **2014**, *347*(3), 200-220.
        [http://dx.doi.org/10.1002/ardp.201300280] [PMID: 24293403]

[4]   Phillips, O.A.; Udo, E.E.; Ali, A.A.; Al-Hassawi, N. Synthesis and antibacterial activity of 5-substituted oxazolidinones. *Bioorg. Med. Chem.,* **2003**, *11*(1), 35-41.
        [http://dx.doi.org/10.1016/S0968-0896(02)00423-6] [PMID: 12467705]

[5]   Almajan, G.L.; Barbuceanu, S.F.; Almajan, E.R.; Draghici, C.; Saramet, G. Synthesis, characterization and antibacterial activity of some triazole Mannich bases carrying diphenylsulfone moieties. *Eur. J. Med. Chem.,* **2009**, *44*(7), 3083-3089.
        [http://dx.doi.org/10.1016/j.ejmech.2008.07.003] [PMID: 18708273]

[6]   Dixit, P.P.; Patil, V.J.; Nair, P.S.; Jain, S.; Sinha, N.; Arora, S.K. Synthesis of 1-[3-(4-benzotria-ol-1/2-yl-3-fluoro-phenyl)-2-oxo-oxazolidin-5-ylmethyl]-3-substituted-thiourea   derivatives   as antituberculosis agents. *Eur. J. Med. Chem.,* **2006**, *41*(3), 423-428.
        [http://dx.doi.org/10.1016/j.ejmech.2005.12.005] [PMID: 16494970]

[7]   Jin, X.; Zheng, C.J.; Song, M.X.; Wu, Y.; Sun, L.P.; Li, Y.J.; Yu, L.J.; Piao, H.R. Synthesis and antimicrobial evaluation of L-phenylalanine-derived C5-substituted rhodanine and chalcone derivatives containing thiobarbituric acid or 2-thioxo-4-thiazolidinone. *Eur. J. Med. Chem.,* **2012**, *56*,

203-209.
[http://dx.doi.org/10.1016/j.ejmech.2012.08.026] [PMID: 22982124]

[8]　Dasgupta, N.; Paul, D.; Dhar Chanda, D.; Ingti, B.; Bhattacharjee, D.; Chakravarty, A.; Bhattacharjee, A. An insight into selection specificity of quinolone resistance determinants within Enterobacteriaceae family. *J. Glob. Antimicrob. Resist.,* **2017**, *10*, 40-46.
[http://dx.doi.org/10.1016/j.jgar.2017.03.010] [PMID: 28619611]

[9]　Raparti, V.; Chitre, T.; Bothara, K.; Kumar, V.; Dangre, S.; Khachane, C.; Gore, S.; Deshmane, B. Novel 4-(morpholin-4-yl)-N′-(arylidene)benzohydrazides: synthesis, antimycobacterial activity and QSAR investigations. *Eur. J. Med. Chem.,* **2009**, *44*(10), 3954-3960.
[http://dx.doi.org/10.1016/j.ejmech.2009.04.023] [PMID: 19464085]

[10]　Woodford, N. Biological counterstrike: antibiotic resistance mechanisms of Gram-positive cocci. *Clin. Microbiol. Infect.,* **2005**, *11* Suppl. 3, 2-21.
[http://dx.doi.org/10.1111/j.1469-0691.2005.01140.x] [PMID: 15811020]

[11]　Mangili, A.; Bica, I.; Snydman, D.R.; Hamer, D.H. Daptomycin-resistant, methicillin-resistant Staphylococcus aureus bacteremia. *Clin. Infect. Dis.,* **2005**, *40*(7), 1058-1060.
[http://dx.doi.org/10.1086/428616] [PMID: 15825002]

[12]　Maruyama, T.; Kano, Y.; Yamamoto, Y.; Kurazono, M.; Iwamatsu, K.; Atsumi, K.; Shitara, E. Synthesis and SAR study of novel 7-(pyridinium-3-yl)-carbonyl imidazo[5,1-b]thiazol-2-yl carbapenems. *Bioorg. Med. Chem.,* **2007**, *15*(1), 392-402.
[http://dx.doi.org/10.1016/j.bmc.2006.09.049] [PMID: 17055731]

[13]　Murray, B.E. Vancomycin-resistant enterococci. *Am. J. Med.,* **1997**, *102*(3), 284-293.
[http://dx.doi.org/10.1016/S0002-9343(99)80270-8] [PMID: 9217598]

[14]　Solomon, V.R.; Hu, C.; Lee, H. Design and synthesis of anti-breast cancer agents from 4-piperazinylquinoline: a hybrid pharmacophore approach. *Bioorg. Med. Chem.,* **2010**, *18*(4), 1563-1572.
[http://dx.doi.org/10.1016/j.bmc.2010.01.001] [PMID: 20106668]

[15]　Hu, C.; Solomon, V.R.; Ulibarri, G.; Lee, H. The efficacy and selectivity of tumor cell killing by Akt inhibitors are substantially increased by chloroquine. *Bioorg. Med. Chem.,* **2008**, *16*(17), 7888-7893.
[http://dx.doi.org/10.1016/j.bmc.2008.07.076] [PMID: 18691894]

[16]　Hu, C.; Raja Solomon, V.; Cano, P.; Lee, H. A 4-aminoquinoline derivative that markedly sensitizes tumor cell killing by Akt inhibitors with a minimum cytotoxicity to non-cancer cells. *Eur. J. Med. Chem.,* **2010**, *45*(2), 705-709.
[http://dx.doi.org/10.1016/j.ejmech.2009.11.017] [PMID: 19945197]

[17]　Ghotaslou, R.; Yekani, M.; Memar, M.Y. The role of efflux pumps in Bacteroides fragilis resistance to antibiotics. *Microbiol. Res.,* **2018**, *210*, 1-5.
[http://dx.doi.org/10.1016/j.micres.2018.02.007] [PMID: 29625653]

[18]　Poirel, L.; Bonnin, R.A.; Nordmann, P. Genetic basis of antibiotic resistance in pathogenic Acinetobacter species. *IUBMB Life,* **2011**, *63*(12), 1061-1067.
[http://dx.doi.org/10.1002/iub.532] [PMID: 21990280]

[19]　Llarrull, L.I.; Testero, S.A.; Fisher, J.F.; Mobashery, S. The future of the β-lactams. *Curr. Opin. Microbiol.,* **2010**, *13*(5), 551-557.
[http://dx.doi.org/10.1016/j.mib.2010.09.008] [PMID: 20888287]

[20]　Laurenzo, D.; Mousa, S.A. Mechanisms of drug resistance in *Mycobacterium tuberculosis* and current status of rapid molecular diagnostic testing. *Acta Trop.,* **2011**, *119*(1), 5-10.
[http://dx.doi.org/10.1016/j.actatropica.2011.04.008] [PMID: 21515239]

[21]　Breidenstein, E.B.; de la Fuente-Núñez, C.; Hancock, R.E. *Pseudomonas aeruginosa*: all roads lead to resistance. *Trends Microbiol.,* **2011**, *19*(8), 419-426.
[http://dx.doi.org/10.1016/j.tim.2011.04.005] [PMID: 21664819]

[22]   Bhardwaj, A.K.; Mohanty, P. Bacterial efflux pumps involved in multidrug resistance and their inhibitors: rejuvinating the antimicrobial chemotherapy. *Recent Pat Antiinfect Drug Discov,* **2012,** *7*(1), 73-89.
[http://dx.doi.org/10.2174/157489112799829710] [PMID: 22353004]

[23]   Payne, D.J.; Gwynn, M.N.; Holmes, D.J.; Pompliano, D.L. Drugs for bad bugs: confronting the challenges of antibacterial discovery. *Nat. Rev. Drug Discov.,* **2007,** *6*(1), 29-40.
[http://dx.doi.org/10.1038/nrd2201] [PMID: 17159923]

[24]   Bagwell, C.L.; Moloney, M.G.; Yaqoob, M. Oxazolomycins: natural product lead structures for novel antibacterials by click fragment conjugation. *Bioorg. Med. Chem. Lett.,* **2010,** *20*(7), 2090-2094.
[http://dx.doi.org/10.1016/j.bmcl.2010.02.066] [PMID: 20223659]

[25]   Guo, X.; Li, Y.L.; Liu, Y.F.; Guo, H.Y.; Wang, Y.C. Synthesis and *in vitro* antibacterial activities of 7-(3-aminopyrrolo[3,4-c]pyrazol-5(*2H,4H,6H*)-yl)quinolone derivatives. *Chin. Chem. Lett.,* **2010,** *21*, 1141-1144.
[http://dx.doi.org/10.1016/j.cclet.2010.06.011]

[26]   Newman, D.J.; Cragg, G.M.; Snader, K.M. Natural products as sources of new drugs over the period 1981-2002. *J. Nat. Prod.,* **2003,** *66*(7), 1022-1037.
[http://dx.doi.org/10.1021/np030096l] [PMID: 12880330]

[27]   Panda, S.S.; Liaqat, S.; Girgis, A.S.; Samir, A.; Hall, C.D.; Katritzky, A.R. Novel antibacterial active quinolone-fluoroquinolone conjugates and 2D-QSAR studies. *Bioorg. Med. Chem. Lett.,* **2015,** *25*(18), 3816-3821.
[http://dx.doi.org/10.1016/j.bmcl.2015.07.077] [PMID: 26253630]

[28]   Basoglu, S.; Yolal, M.; Demirbas, A.; Bektas, H.; Abbasoglu, R.; Demirbas, N. Synthesis of linezolid-like molecules and evaluation of their antimicrobial activities. *Turk. J. Chem.,* **2012,** *36*, 37-53.

[29]   Ceylan, S.; Bektas, H.; Bayrak, H.; Demirbas, N.; Alpay-Karaoglu, S.; Ulker, S. Syntheses and biological activities of new hybrid molecules containing different heterocyclic moieties. *Arch. Pharm. (Weinheim),* **2013,** *346*(10), 743-756.
[http://dx.doi.org/10.1002/ardp.201300161] [PMID: 24038519]

[30]   Shaveta, ; Mishra, S.; Singh, P. Hybrid molecules: The privileged scaffolds for various pharmaceuticals. *Eur. J. Med. Chem.,* **2016,** *124*, 500-536.
[http://dx.doi.org/10.1016/j.ejmech.2016.08.039] [PMID: 27598238]

[31]   Zhang, H.; Tian, Y.; Kang, D.; Huo, Z.; Zhou, Z.; Liu, H.; De Clercq, E.; Pannecouque, C.; Zhan, P.; Liu, X. Discovery of uracil-bearing DAPYs derivatives as novel HIV-1 NNRTIs *via* crystallographic overlay-based molecular hybridization. *Eur. J. Med. Chem.,* **2017,** *130*, 209-222.
[http://dx.doi.org/10.1016/j.ejmech.2017.02.047] [PMID: 28254696]

[32]   Vandekerckhove, S.; Desmet, T.; Tran, H.G.; de Kock, C.; Smith, P.J.; Chibale, K.; D'hooghe, M. Synthesis of halogenated 4-quinolones and evaluation of their antiplasmodial activity. *Bioorg. Med. Chem. Lett.,* **2014,** *24*(4), 1214-1217.
[http://dx.doi.org/10.1016/j.bmcl.2013.12.067] [PMID: 24468411]

[33]   Zhang, G.F.; Liu, X.; Zhang, S.; Pan, B.; Liu, M.L. Ciprofloxacin derivatives and their antibacterial activities. *Eur. J. Med. Chem.,* **2018,** *146*, 599-612.
[http://dx.doi.org/10.1016/j.ejmech.2018.01.078] [PMID: 29407984]

[34]   Chugunova, E.; Akylbekov, N.; Bulatova, A.; Gavrilov, N.; Voloshina, A.; Kulik, N.; Zobov, V.; Dobrynin, A.; Syakaev, V.; Burilov, A. Synthesis and biological evaluation of novel structural hybrids of benzofuroxan derivatives and fluoroquinolones. *Eur. J. Med. Chem.,* **2016,** *116*, 165-172.
[http://dx.doi.org/10.1016/j.ejmech.2016.03.086] [PMID: 27061980]

[35]   Sabatini, S.; Gosetto, F.; Manfroni, G.; Tabarrini, O.; Kaatz, G.W.; Patel, D.; Cecchetti, V. Evolution from a natural flavones nucleus to obtain 2-(4-Propoxyphenyl)quinoline derivatives as potent inhibitors of the S. aureus NorA efflux pump. *J. Med. Chem.,* **2011,** *54*(16), 5722-5736.

[http://dx.doi.org/10.1021/jm200370y] [PMID: 21751791]

[36]   Srinivasan, S.; Beema Shafreen, R.M.; Nithyanand, P.; Manisankar, P.; Pandian, S.K. Synthesis and *in vitro* antimicrobial evaluation of novel fluoroquinolone derivatives. *Eur. J. Med. Chem.,* **2010**, *45*(12), 6101-6105.
[http://dx.doi.org/10.1016/j.ejmech.2010.09.036] [PMID: 20933306]

[37]   Charushin, V.N.; Mochulskaya, N.N. Antipin, Kotovskaya, S.K.; Nosova, E.V.; Ezhikova, M.A.; Kodess, M.I.; Kravchenko, M.A. Synthesis and antimycobacterial evaluation of new (2-oxo-2*H*-chromen-3- yl) substituted fluoroquinolones. *J. Fluor. Chem.,* **2018**, *208*, 15-23.
[http://dx.doi.org/10.1016/j.jfluchem.2018.01.007]

[38]   Xiao, Z.P.; Wang, X.D.; Wang, P.F.; Zhou, Y.; Zhang, J.W.; Zhang, L.; Zhou, J.; Zhou, S.S.; Ouyang, H.; Lin, X.Y.; Mustapa, M.; Reyinbaike, A.; Zhu, H.L. Design, synthesis, and evaluation of novel fluoroquinolone-flavonoid hybrids as potent antibiotics against drug-resistant microorganisms. *Eur. J. Med. Chem.,* **2014**, *80*, 92-100.
[http://dx.doi.org/10.1016/j.ejmech.2014.04.037] [PMID: 24769347]

[39]   Sutera, V.; Hoarau, G.; Renesto, P.; Caspar, Y.; Maurin, M. *In vitro* and *in vivo* evaluation of fluoroquinolone resistance associated with DNA gyrase mutations in Francisella tularensis, including in tularaemia patients with treatment failure. *Int. J. Antimicrob. Agents,* **2017**, *50*(3), 377-383.
[http://dx.doi.org/10.1016/j.ijantimicag.2017.03.022] [PMID: 28689870]

[40]   Zhang, G.F.; Zhang, S.; Pan, B.; Liu, X.; Feng, L.S. 4-Quinolone derivatives and their activities against Gram positive pathogens. *Eur. J. Med. Chem.,* **2018**, *143*, 710-723.
[http://dx.doi.org/10.1016/j.ejmech.2017.11.082] [PMID: 29220792]

[41]   Palaska, E. *Farmasötik Kimya, Hacettepe University Press., Second Eddition*; ANKARA, **2004**.

[42]   Nagaraja, V.; Godbole, A.A.; Henderson, S.R.; Maxwell, A. DNA topoisomerase I and DNA gyrase as targets for TB therapy. *Drug Discov. Today,* **2017**, *22*(3), 510-518.
[http://dx.doi.org/10.1016/j.drudis.2016.11.006] [PMID: 27856347]

[43]   Mao, T.Q.; He, Q.Q.; Wan, Z.Y.; Chen, W.X.; Chen, F.E.; Tang, G.F.; De Clercq, E.; Daelemans, D.; Pannecouque, C. Anti-HIV diarylpyrimidine-quinolone hybrids and their mode of action. *Bioorg. Med. Chem.,* **2015**, *23*(13), 3860-3868.
[http://dx.doi.org/10.1016/j.bmc.2015.03.037] [PMID: 25907370]

[44]   Domagala, J.M.; Heifetz, C.L.; Hutt, M.P.; Mich, T.F.; Nichols, J.B.; Solomon, M.; Worth, D.F. 1-Substituted   7-[3-[(ethylamino)methyl]-1-pyrrolidinyl]-6,8-   difluoro-1,4-dihydro-4--xo-3-quinolinecarboxylic acids. New quantitative structure-activity relationships at N1 for the quinolone antibacterials. *J. Med. Chem.,* **1988**, *31*(5), 991-1001.
[http://dx.doi.org/10.1021/jm00400a017] [PMID: 2834557]

[45]   Itoh, K.; Kuramoto, Y.; Amano, H.; Kazamori, D.; Yazaki, A. Discovery of WQ-3810: Design, synthesis, and evaluation of 7-(3-alkylaminoazetidin-1-yl)fluoro-quinolones as orally active antibacterial agents. *Eur. J. Med. Chem.,* **2015**, *103*, 354-360.
[http://dx.doi.org/10.1016/j.ejmech.2015.08.015] [PMID: 26363871]

[46]   Kuramoto, Y.; Ohshita, Y.; Yoshida, J.; Yazaki, A.; Shiro, M.; Koike, T. A novel antibacterial 8-chloroquinolone with a distorted orientation of the N1-(5-amino-2,4-difluorophenyl) group. *J. Med. Chem.,* **2003**, *46*(10), 1905-1917.
[http://dx.doi.org/10.1021/jm0205090] [PMID: 12723953]

[47]   Kant, R.; Singh, V.; Nath, G.; Awasthi, S.K.; Agarwal, A. Design, synthesis and biological evaluation of ciprofloxacin tethered bis-1,2,3-triazole conjugates as potent antibacterial agents. *Eur. J. Med. Chem.,* **2016**, *124*, 218-228.
[http://dx.doi.org/10.1016/j.ejmech.2016.08.031] [PMID: 27592391]

[48]   Srivastava, B.K.; Solanki, M.; Mishra, B.; Soni, R.; Jayadev, S.; Valani, D.; Jain, M.; Patel, P.R. Synthesis and antibacterial activity of 4,5,6,7-tetrahydro-thieno[3,2-c]pyridine quinolones. *Bioorg. Med. Chem. Lett.,* **2007**, *17*(7), 1924-1929.

[http://dx.doi.org/10.1016/j.bmcl.2007.01.038] [PMID: 17276057]

[49]   Suthar, S.K.; Jaiswal, V.; Lohan, S.; Bansal, S.; Chaudhary, A.; Tiwari, A.; Alex, A.T.; Joesph, A. Novel quinolone substituted thiazolidin-4-ones as anti-inflammatory, anticancer agents: design, synthesis and biological screening. *Eur. J. Med. Chem.,* **2013**, *63*, 589-602.
[http://dx.doi.org/10.1016/j.ejmech.2013.03.011] [PMID: 23548704]

[50]   Hu, Y.Q.; Zhang, S.; Xu, Z.; Lv, Z.S.; Liu, M.L.; Feng, L.S. 4-Quinolone hybrids and their antibacterial activities. *Eur. J. Med. Chem.,* **2017**, *141*, 335-345.
[http://dx.doi.org/10.1016/j.ejmech.2017.09.050] [PMID: 29031077]

[51]   Lv, K.; Liu, M.L.; Feng, L.S.; Sun, L.Y.; Sun, Y.X.; Wei, Z.Q.; Guo, H.Q. Synthesis and antibacterial activity of naphthyridone derivatives containing mono/difluoro-methyloxime pyrrolidine scaffolds. *Eur. J. Med. Chem.,* **2012**, *47*(1), 619-625.
[http://dx.doi.org/10.1016/j.ejmech.2011.10.048] [PMID: 22079379]

[52]   Choi, D.R.; Shin, J.H.; Yang, J.; Yoon, S.H.; Jung, Y.H. Syntheses and biological evaluation of new fluoroquinolone antibacterials containing chiral oxiimino pyrrolidine. *Bioorg. Med. Chem. Lett.,* **2004**, *14*(5), 1273-1277.
[http://dx.doi.org/10.1016/j.bmcl.2003.12.044] [PMID: 14980680]

[53]   Lv, K.; Wu, J.; Wang, J.; Liu, M.; Wei, Z.; Cao, J.; Sun, Y.; Guo, H. Synthesis and *in vitro* antibacterial activity of quinolone/naphthyridone derivatives containing 3-alkoxyimino-4-(methyl)aminopiperidine scaffolds. *Bioorg. Med. Chem. Lett.,* **2013**, *23*(6), 1754-1759.
[http://dx.doi.org/10.1016/j.bmcl.2013.01.048] [PMID: 23402878]

[54]   Rusu, A.; Hancu, G.; Munteanu, A.C.; Uivarosi, V. Development perspectives of silver complexes with antibacterial quinolones: Successful or not? *J. Organomet. Chem.,* **2017**, *839*, 19-30.
[http://dx.doi.org/10.1016/j.jorganchem.2017.02.023]

[55]   Ge, R.; Zhao, Q.; Xie, Z.; Lu, L.; Guo, Q.; Li, Z.; Zhao, L. Synthesis and biological evaluation of 6-fluoro-3-phenyl-7-piperazinyl quinolone derivatives as potential topoisomerase I inhibitors. *Eur. J. Med. Chem.,* **2016**, *122*, 465-474.
[http://dx.doi.org/10.1016/j.ejmech.2016.06.054] [PMID: 27416553]

[56]   Foroumadi, A.; Mansouri, S.; Kiani, Z.; Rahmani, A. Synthesis and *in vitro* antibacterial evaluation of *N*-[5-(5-nitro-2-thienyl)-1,3,4-thiadiazole-2-yl] piperazinyl quinolones. *Eur. J. Med. Chem.,* **2003**, *38*(9), 851-854.
[http://dx.doi.org/10.1016/S0223-5234(03)00148-X] [PMID: 14561484]

[57]   Foroumadi, A.; Ghodsi, S.; Emami, S.; Najjari, S.; Samadi, N.; Faramarzi, M.A.; Beikmohammadi, L.; Shirazid, F.H.; Shafiee, A. Synthesis and antibacterial activity of new fluoroquinolones containing a substituted *N*-(phenethyl)piperazine moiety. *Eur. J. Med. Chem.,* **2007**, *42*, 985-992.
[http://dx.doi.org/10.1016/j.ejmech.2006.12.034] [PMID: 17316916]

[58]   Mentese, M.; Demirbas, N.; Mermer, A.; Demirci, S.; Demirbas, A.; Ayaz, F.A. Novel Azole-Functionalized fluoroquinolone hybrids: design, conventional and microwave irradiated synthesis, evaluation as antibacterial and antioxidant agents. *Lett. Drug Des. Discov.,* **2018**, *15*, 46-64.
[http://dx.doi.org/10.2174/1570180814666170823163540]

[59]   Foroumadi, A.; Soltani, F.; Moshafi, M.H.; Ashraf-Askari, R. Synthesis and *in vitro* antibacterial activity of some *N*-(5-aryl-1,3,4-thiadiazole-2-yl)piperazinyl quinolone derivatives. *Farmaco,* **2003**, *58*(10), 1023-1028.
[http://dx.doi.org/10.1016/S0014-827X(03)00191-5] [PMID: 14505733]

[60]   Foroumadi, A.; Ashraf-Askari, R.; Moshafi, M.H.; Emami, S.; Zeynali, A. Synthesis and *in vitro* antibacterial activity of *N*-[5-(5-nitro-2-furyl)-1,3,4-thiadiazole-2-yl] piperazinyl quinolone derivatives. *Pharmazie,* **2003**, *58*(6), 432-433.
[PMID: 12857013]

[61]   Foroumadi, A.; Mansouri, S.; Emami, S.; Mirzaei, J.; Sorkhi, M.; Saeid-Adeli, N.; Shafiee, A. Synthesis and antibacterial activity of nitroaryl thiadiazole-levofloxacin hybrids. *Arch. Pharm.*

(Weinheim), **2006**, *339*(11), 621-624.
[http://dx.doi.org/10.1002/ardp.200600108] [PMID: 17036368]

[62]   Foroumadi, A.; Emami, S.; Hassanzadeh, A.; Rajaee, M.; Sokhanvar, K.; Moshafi, M.H.; Shafiee, A. Synthesis and antibacterial activity of N-(5-benzylthio-1,3,4-thiadiazol-2-yl) and *N*-(5-benzylsulfon-l-1,3,4-thiadiazol-2-yl)piperazinyl quinolone derivatives. *Bioorg. Med. Chem. Lett.,* **2005**, *15*(20), 4488-4492.
       [http://dx.doi.org/10.1016/j.bmcl.2005.07.016] [PMID: 16105736]

[63]   Feng, L.S.; Liu, M.L.; Zhang, S.; Chai, Y.; Wang, B.; Zhang, Y.B.; Lv, K.; Guan, Y.; Guo, H.Y.; Xiao, C.L. Synthesis and *in vitro* antimycobacterial activity of 8-OCH$_3$ ciprofloxacin methylene and ethylene isatin derivatives. *Eur. J. Med. Chem.,* **2011**, *46*(1), 341-348.
       [http://dx.doi.org/10.1016/j.ejmech.2010.11.023] [PMID: 21146257]

[64]   Foroumadi, A.; Emami, S.; Mehni, M.; Moshafi, M.H.; Shafiee, A. Synthesis and antibacterial activity of *N*-[2-(5-bromothiophen-2-yl)-2-oxoethyl] and *N*-[(2-5-bromothiophen-2-yl)-2-oximinoethyl] derivatives of piperazinyl quinolones. *Bioorg. Med. Chem. Lett.,* **2005**, *15*(20), 4536-4539.
       [http://dx.doi.org/10.1016/j.bmcl.2005.07.005] [PMID: 16115766]

[65]   Rabbani, M.G.; Islam, M.R.; Ahmad, M.; Hossion, A.M.L. Synthesis of some NH Derivatives of ciprofloxacin as antibacterial and antifungal agents. *Bangladesh J. Pharmacol.,* **2011**, *6*, 8-13.
       [http://dx.doi.org/10.3329/bjp.v6i1.7720]

[66]   Faidallah, H.M.; Girgis, A.S.; Tiwari, A.D.; Honkanadavar, H.H.; Thomas, S.J.; Samir, A.; Kalmouch, A.; Alamry, K.A.; Khan, K.A.; Ibrahim, T.S.; Al-Mahmoudy, A.M.M.; Asiri, A.M.; Panda, S.S. Synthesis, antibacterial properties and 2D-QSAR studies of quinolone-triazole conjugates. *Eur. J. Med. Chem.,* **2018**, *143*, 1524-1534.
       [http://dx.doi.org/10.1016/j.ejmech.2017.10.042] [PMID: 29126731]

[67]   Cheng, Y.; Avula, S.R.; Gao, W.W.; Addla, D.; Tangadanchu, V.K.R.; Zhang, L.; Lin, J.M.; Zhou, C.H. Multi-targeting exploration of new 2-aminothiazolyl quinolones: Synthesis, antimicrobial evaluation, interaction with DNA, combination with topoisomerase IV and penetrability into cells. *Eur. J. Med. Chem.,* **2016**, *124*, 935-945.
       [http://dx.doi.org/10.1016/j.ejmech.2016.10.011] [PMID: 27769037]

[68]   Cormier, R.; Burda, W.N.; Harrington, L.; Edlinger, J.; Kodigepalli, K.M.; Thomas, J.; Kapolka, R.; Roma, G.; Anderson, B.E.; Turos, E.; Shaw, L.N. Studies on the antimicrobial properties of *N*-acylated ciprofloxacins. *Bioorg. Med. Chem. Lett.,* **2012**, *22*(20), 6513-6520.
       [http://dx.doi.org/10.1016/j.bmcl.2012.05.026] [PMID: 22995622]

[69]   McPherson, J.C., III; Runner, R.; Buxton, T.B.; Hartmann, J.F.; Farcasiu, D.; Bereczki, I.; Roth, E.; Tollas, S.; Ostorházi, E.; Rozgonyi, F.; Herczegh, P. Synthesis of osteotropic hydroxybisphosphonate derivatives of fluoroquinolone antibacterials. *Eur. J. Med. Chem.,* **2012**, *47*(1), 615-618.
       [http://dx.doi.org/10.1016/j.ejmech.2011.10.049] [PMID: 22093760]

[70]   Wang, S.; Jia, X.D.; Liu, M.L.; Lu, Y.; Guo, H.Y. Synthesis, antimycobacterial and antibacterial activity of ciprofloxacin derivatives containing a N-substituted benzyl moiety. *Bioorg. Med. Chem. Lett.,* **2012**, *22*(18), 5971-5975.
       [http://dx.doi.org/10.1016/j.bmcl.2012.07.040] [PMID: 22884110]

[71]   Dheer, D.; Singh, V.; Shankar, R. Medicinal attributes of 1,2,3-triazoles: Current developments. *Bioorg. Chem.,* **2017**, *71*, 30-54.
       [http://dx.doi.org/10.1016/j.bioorg.2017.01.010] [PMID: 28126288]

[72]   Hayashi, N.; Nakata, Y.; Yazaki, A. New findings on the structure-phototoxicity relationship and photostability of fluoroquinolones with various substituents at position 1. *Antimicrob. Agents Chemother.,* **2004**, *48*(3), 799-803.
       [http://dx.doi.org/10.1128/AAC.48.3.799-803.2004] [PMID: 14982767]

[73]   Shavit, M.; Pokrovskaya, V.; Belakhov, V.; Baasov, T. Covalently linked kanamycin - Ciprofloxacin hybrid antibiotics as a tool to fight bacterial resistance. *Bioorg. Med. Chem.,* **2017**, *25*(11), 2917-2925.

[http://dx.doi.org/10.1016/j.bmc.2017.02.068] [PMID: 28343755]

[74]   Jegatheeswaran, S.; Selvam, S.; Ramkumar, V.S.; Sundrarajan, M. Facile green synthesis of silver doped fluor-hydroxyapatite/-cyclodextrin nanocomposite in the dual acting fluorine-containing ionic liquid medium for bone substitute applications. *Appl. Surf. Sci.,* **2016**, *371*, 468-478.
[http://dx.doi.org/10.1016/j.apsusc.2016.03.007]

[75]   Panda, S.S.; Jain, S.C. New trifluoromethyl quinolone derivatives: synthesis and investigation of antimicrobial properties. *Bioorg. Med. Chem. Lett.,* **2013**, *23*(11), 3225-3229.
[http://dx.doi.org/10.1016/j.bmcl.2013.03.120] [PMID: 23611733]

[76]   German, N.; Wei, P.; Kaatz, G.W.; Kerns, R.J. Synthesis and evaluation of fluoroquinolone derivatives as substrate-based inhibitors of bacterial efflux pumps. *Eur. J. Med. Chem.,* **2008**, *43*(11), 2453-2463.
[http://dx.doi.org/10.1016/j.ejmech.2008.01.042] [PMID: 18358571]

[77]   Kanagaratnam, R.; Sheikh, R.; Alharbi, F.; Kwon, D.H. An efflux pump (MexAB-OprM) of Pseudomonas aeruginosa is associated with antibacterial activity of Epigallocatechin-3-gallate (EGCG). *Phytomedicine,* **2017**, *36*, 194-200.
[http://dx.doi.org/10.1016/j.phymed.2017.10.010] [PMID: 29157815]

[78]   Kristiansen, J.E.; Hendricks, O.; Delvin, T.; Butterworth, T.S.; Aagaard, L.; Christensen, J.B.; Flores, V.C.; Keyzer, H. Reversal of resistance in microorganisms by help of non-antibiotics. *J. Antimicrob. Chemother.,* **2007**, *59*(6), 1271-1279.
[http://dx.doi.org/10.1093/jac/dkm071] [PMID: 17403708]

[79]   Miladi, H.; Zmantar, T.; Chaabouni, Y.; Fedhila, K.; Bakhrouf, A.; Mahdouani, K.; Chaieb, K. Antibacterial and efflux pump inhibitors of thymol and carvacrol against food-borne pathogens. *Microb. Pathog.,* **2016**, *99*, 95-100.
[http://dx.doi.org/10.1016/j.micpath.2016.08.008] [PMID: 27521228]

[80]   Lomovskaya, O.; Bostian, K.A. Practical applications and feasibility of efflux pump inhibitors in the clinic--a vision for applied use. *Biochem. Pharmacol.,* **2006**, *71*(7), 910-918.
[http://dx.doi.org/10.1016/j.bcp.2005.12.008] [PMID: 16427026]

[81]   Sheppard, J.G.; Long, T.E. Allicin-inspired thiolated fluoroquinolones as antibacterials against ESKAPE pathogens. *Bioorg. Med. Chem. Lett.,* **2016**, *26*(22), 5545-5549.
[http://dx.doi.org/10.1016/j.bmcl.2016.10.002] [PMID: 27756563]

[82]   Frigola, J.; Parés, J.; Corbera, J.; Vañó, D.; Mercè, R.; Torrens, A.; Más, J.; Valentí, E. 7-Azetidinylquinolones as antibacterial agents. Synthesis and structure-activity relationships. *J. Med. Chem.,* **1993**, *36*(7), 801-810.
[http://dx.doi.org/10.1021/jm00059a002] [PMID: 8464033]

[83]   Frigola, J.; Torrens, A.; Castrillo, J.A.; Mas, J.; Vañó, D.; Berrocal, J.M.; Calvet, C.; Salgado, L.; Redondo, J.; García-Granda, S.; Valenti, E.; Quintana, J.R. 7-azetidinylquinolones as antibacterial agents. 2. Synthesis and biological activity of 7-(2,3-disubstituted-1-azetidinyl)-4-oxoquinoline- and -1,8-naphthyridine-3-carboxylic acids. Properties and structure-activity relationships of quinolones with an azetidine moiety. *J. Med. Chem.,* **1994**, *37*(24), 4195-4210.
[http://dx.doi.org/10.1021/jm00050a016] [PMID: 7990118]

[84]   Frigola, J.; Vañó, D.; Torrens, A.; Gómez-Gomar, A.; Ortega, E.; García-Granda, S. 7-Azetidinylquinolones as antibacterial agents. 3. Synthesis, properties and structure-activity relationships of the stereoisomers containing a 7-(3-amino-2-methyl-1-azetidinyl) moiety. *J. Med. Chem.,* **1995**, *38*(7), 1203-1215.
[http://dx.doi.org/10.1021/jm00007a017] [PMID: 7707323]

[85]   Ikee, Y.; Hashimoto, K.; Nakashima, M.; Hayashi, K.; Sano, S.; Shiro, M.; Nagao, Y. Synthesis and antibacterial activities of new quinolone derivatives utilizing 1-azabicyclo[1.1.0]butane. *Bioorg. Med. Chem. Lett.,* **2007**, *17*(4), 942-945.
[http://dx.doi.org/10.1016/j.bmcl.2006.11.048] [PMID: 17157008]

[86]   Güzeldemirci, N.U.; Küçükbasmaci, O. Synthesis and antimicrobial activity evaluation of new 1,2,4-triazoles and 1,3,4-thiadiazoles bearing imidazo[2,1-b]thiazole moiety. *Eur. J. Med. Chem.,* **2010,** *45*(1), 63-68.
[http://dx.doi.org/10.1016/j.ejmech.2009.09.024] [PMID: 19939519]

[87]   Bayrak, H.; Demirbas, A.; Demirbas, N.; Karaoglu, S.A. Synthesis of some new 1,2,4-triazoles starting from isonicotinic acid hydrazide and evaluation of their antimicrobial activities. *Eur. J. Med. Chem.,* **2009,** *44*(11), 4362-4366.
[http://dx.doi.org/10.1016/j.ejmech.2009.05.022] [PMID: 19647352]

[88]   Tozkoparan, B.; Küpeli, E.; Yeşilada, E.; Ertan, M. Preparation of 5-aryl-3-alkylthio-1,2,4-triazoles and corresponding sulfones with antiinflammatory-analgesic activity. *Bioorg. Med. Chem.,* **2007,** *15*(4), 1808-1814.
[http://dx.doi.org/10.1016/j.bmc.2006.11.029] [PMID: 17166724]

[89]   Stefanska, J.; Szulczyk, D.; Koziol, A.E.; Miroslaw, B.; Kedzierska, E.; Fidecka, S.; Busonera, B.; Sanna, G.; Giliberti, G.; La Colla, P.; Struga, M. Disubstituted thiourea derivatives and their activity on CNS: synthesis and biological evaluation. *Eur. J. Med. Chem.,* **2012,** *55*, 205-213.
[http://dx.doi.org/10.1016/j.ejmech.2012.07.020] [PMID: 22884523]

[90]   Suresh Kumar, G.V.; Rajendraprasad, Y.; Mallikarjuna, B.P.; Chandrashekar, S.M.; Kistayya, C. Synthesis of some novel 2-substituted-5-[isopropylthiazole] clubbed 1,2,4-triazole and 1,3,4-oxadiazoles as potential antimicrobial and antitubercular agents. *Eur. J. Med. Chem.,* **2010,** *45*(5), 2063-2074.
[http://dx.doi.org/10.1016/j.ejmech.2010.01.045] [PMID: 20149496]

[91]   Husain, A.; Rashid, M.; Shaharyar, M.; Siddiqui, A.A.; Mishra, R. Benzimidazole clubbed with triazolo-thiadiazoles and triazolo-thiadiazines: new anticancer agents. *Eur. J. Med. Chem.,* **2013,** *62*, 785-798.
[http://dx.doi.org/10.1016/j.ejmech.2012.07.011] [PMID: 23333063]

[92]   Demirbaş, N.; Ugurluoglu, R.; Demirbaş, A. Synthesis of 3-alkyl(aryl)-4-alkylidenamino-4,5-diydro-1H-1,2,4-triazol-5-ones   and   3-alkyl-4-alkylamino-4,5-dihydro-1H-1,2,4-triazol-5-ones   as antitumor agents. *Bioorg. Med. Chem.,* **2002,** *10*(12), 3717-3723.
[http://dx.doi.org/10.1016/S0968-0896(02)00420-0] [PMID: 12413828]

[93]   Demirbas, N.; Karaoglu, S.A.; Demirbas, A.; Sancak, K. Synthesis and antimicrobial activities of some new 1-(5-phenylamino-[1,3,4]thiadiazol-2-yl)methyl-5-oxo-[1,2,4]triazole and 1-(4-pheny--5-thioxo-[1,2,4]triazol-3-yl)methyl-5-oxo-{1,2,4]triazole derivatives. *Eur. J. Med. Chem.,* **2004,** *39*, 793-804.
[http://dx.doi.org/10.1016/j.ejmech.2004.06.007] [PMID: 15337292]

[94]   Johnson, E.; Espinel-Ingroff, A.; Szekely, A.; Hockey, H.; Troke, P. Activity of voriconazole, itraconazole, fluconazole and amphotericin B *in vitro* against 1763 yeasts from 472 patients in the voriconazole phase III clinical studies. *Int. J. Antimicrob. Agents,* **2008,** *32*(6), 511-514.
[http://dx.doi.org/10.1016/j.ijantimicag.2008.05.023] [PMID: 18790613]

[95]   Nitenberg, G.; Raynard, B.; Bichisao, E.; Pozzi, P.; Toffolatti, L. Nutritional support of the cancer patient: issues and dilemmas. *Crit. Rev. Oncol. Hematol.,* **2000,** *34*(3), 137-168.
[http://dx.doi.org/10.1016/S1040-8428(00)00048-2] [PMID: 10838261]

[96]   Demirbas, A.; Ceylan, S.; Demirbas, N. Synthesis of some new five membered heterocycles, a facile synthesis of oxazolidinones. *J. Heterocycl. Chem.,* **2007,** *44*, 1271-1280.
[http://dx.doi.org/10.1002/jhet.5570440608]

[97]   Verma, A.; Saraf, S.K. 4-thiazolidinone--a biologically active scaffold. *Eur. J. Med. Chem.,* **2008,** *43*(5), 897-905.
[http://dx.doi.org/10.1016/j.ejmech.2007.07.017] [PMID: 17870209]

[98]   Demirbas, N.; Demirbas, A.; Alpay-Karaoglu, S.; Çelik, E. Synthesis and antimicrobial activities of some new [1,2,4]triazolo[3,4-b][1,3,4]thiadiazoles and [1,2,4]triazolo[3,4-b] [1,3,4]thiadiazines.

ARKIVOC, **2005**, *i,* 75-91.

[99]   Menteşe, M.; Beriş, F.S.; Demirbaş, N. Synthesis of Some New Ciprofloxacin Hybrids as Potential Antimicrobial Agents. *J. Heterocycl. Chem.,* **2017**, *54,* 2996-3007.
[http://dx.doi.org/10.1002/jhet.2907]

[100]   Demirci, S.; Mermer, A.; Ak, G.; Aksakal, F.; Colak, N.; Demirbas, A.; Ayaz, F.A.; Demirbas, N. Conventional and Microwave-assisted Total Synthesis, Antioxidant Capacity, Biological Activity, and Molecular Docking Studies of New Hybrid Compounds. *J. Heterocycl. Chem.,* **2017**, *54,* 1785-1805.
[http://dx.doi.org/10.1002/jhet.2760]

[101]   Mentese, M.; Demirci, S.; Ozdemir, S.B.; Demirbas, A.; Ulker, S.; Demirbas, N. Microwave Assisted Synthesis and Antimicrobial Activity Evaluation of New Heterofunctionalized Norfloxacine Derivatives. *Lett. Drug Des. Discov.,* **2016**, *13,* 1076-1090.
[http://dx.doi.org/10.2174/1570180813666160712222922]

[102]   Basoglu, S.; Ulker, S.; Alpay-Karaoglu, S.; Demirbas, N. Microwave-assisted synthesis of some hybrid molecules containing penicillanic acid or cephalosporanic acid moieties and investigation of their biological activities. *Med. Chem. Res.,* **2014**, *23,* 3128-3143.
[http://dx.doi.org/10.1007/s00044-013-0898-4] [PMID: 24719549]

[103]   Zhang, J.; Wang, S.; Wei, Q.; Guo, Q.; Bai, Y.; Yang, S.; Song, F.; Zhang, L.; Lei, X. Synthesis and biological evaluation of Aspergillomarasmine A derivatives as novel NDM-1 inhibitor to overcome antibiotics resistance. *Bioorg. Med. Chem.,* **2017**, *25*(19), 5133-5141.
[http://dx.doi.org/10.1016/j.bmc.2017.07.025] [PMID: 28784300]

[104]   Yolal, M.; Basoglu, S.; Bektas, H.; Demirci, S.; Alpay-Karaoglu, S.; Demirbas, A. Synthesis of eperezolid-like molecules and evaluation of their antimicrobial activities. *Bioorg. Khim.,* **2012**, *38*(5), 610-620.
[PMID: 23342495]

[105]   Mermer, A.; Demirci, S.; Ozdemir, S.B.; Demirbas, A.; Ulker, S.; Ayaz, F.A.; Aksakal, F.; Demirbas, N. Conventional and microwave irradiated synthesis, biological activity evaluation and molecular docking studies of highly substituted piperazine-azole hybrids. *Chin. Chem. Lett.,* **2017**, *28,* 995-1005.
[http://dx.doi.org/10.1016/j.cclet.2016.12.012]

[106]   Roman, G. Mannich bases in medicinal chemistry and drug design. *Eur. J. Med. Chem.,* **2015**, *89,* 743-816.
[http://dx.doi.org/10.1016/j.ejmech.2014.10.076] [PMID: 25462280]

[107]   Plech, T.; Wujec, M.; Kosikowska, U.; Malm, A.; Rajtar, B.; Polz-Dacewicz, M. Synthesis and *in vitro* activity of 1,2,4-triazole-ciprofloxacin hybrids against drug-susceptible and drug-resistant bacteria. *Eur. J. Med. Chem.,* **2013**, *60,* 128-134.
[http://dx.doi.org/10.1016/j.ejmech.2012.11.040] [PMID: 23287058]

[108]   Plech, T.; Kaproń, B.; Paneth, A.; Kosikowska, U.; Malm, A.; Strzelczyk, A.; Stączek, P.; Świątek, Ł.; Rajtar, B.; Polz-Dacewicz, M. Search for factors affecting antibacterial activity and toxicity of 1,2,4-triazole-ciprofloxacin hybrids. *Eur. J. Med. Chem.,* **2015**, *97,* 94-103.
[http://dx.doi.org/10.1016/j.ejmech.2015.04.058] [PMID: 25951434]

[109]   Wang, Y.; Damu, G.L.V.; Lv, J.S.; Geng, R.X.; Yang, D.C.; Zhou, C.H. Design, synthesis and evaluation of clinafloxacin triazole hybrids as a new type of antibacterial and antifungal agents. *Bioorg. Med. Chem. Lett.,* **2012**, *22*(17), 5363-5366.
[http://dx.doi.org/10.1016/j.bmcl.2012.07.064] [PMID: 22884108]

[110]   Gao, L.Z.; Xie, Y.S.; Li, T.; Huang, W.L.; Hu, G.Q. Synthesis and antibacterial activity of novel [1,2,4]triazolo[3,4-h][1,8]naphthyridine-7-carboxylic acid derivatives. *Chin. Chem. Lett.,* **2015**, *26,* 149-151.
[http://dx.doi.org/10.1016/j.cclet.2014.09.017]

[111]   Carta, A.; Palomba, M.; Briguglio, I.; Corona, P.; Piras, S.; Jabes, D.; Guglierame, P.; Molicotti, P.; Zanetti, S. Synthesis and anti-mycobacterial activities of triazoloquinolones. *Eur. J. Med. Chem.,*

2011, *46*(1), 320-326.
[http://dx.doi.org/10.1016/j.ejmech.2010.11.020] [PMID: 21145625]

[112]   Aggarwal, N.; Kumar, R.; Dureja, P.; Khurana, J.M. Synthesis, antimicrobial evaluation and QSAR analysis of novel nalidixic acid based 1,2,4-triazole derivatives. *Eur. J. Med. Chem.,* **2011**, *46*(9), 4089-4099.
[http://dx.doi.org/10.1016/j.ejmech.2011.06.009] [PMID: 21752498]

[113]   Kumar, D.; Patel, G.; Chavers, A.K.; Chang, K.H.; Shah, K. Synthesis of novel 1,2,4-oxadiazoles and analogues as potential anticancer agents. *Eur. J. Med. Chem.,* **2011**, *46*(7), 3085-3092.
[http://dx.doi.org/10.1016/j.ejmech.2011.03.031] [PMID: 21481985]

[114]   Chandrakantha, B.; Shetty, P.; Nambiyar, V.; Isloor, N.; Isloor, A.M. Synthesis, characterization and biological activity of some new 1,3,4-oxadiazole bearing 2-flouro-4-methoxy phenyl moiety. *Eur. J. Med. Chem.,* **2010**, *45*(3), 1206-1210.
[http://dx.doi.org/10.1016/j.ejmech.2009.11.046] [PMID: 20004043]

[115]   Jha, K.K.; Samad, A.; Kumar, Y.; Shaharyar, M.; Khosa, R.L.; Jain, J.; Kumar, V.; Singh, P. Design, synthesis and biological evaluation of 1,3,4-oxadiazole derivatives. *Eur. J. Med. Chem.,* **2010**, *45*(11), 4963-4967.
[http://dx.doi.org/10.1016/j.ejmech.2010.08.003] [PMID: 20817328]

[116]   Li, Y.; Luo, Y.; Hu, Y.; Zhu, D.D.; Zhang, S.; Liu, Z.J.; Gong, H.B.; Zhu, H.L. Design, synthesis and antimicrobial activities of nitroimidazole derivatives containing 1,3,4-oxadiazole scaffold as FabH inhibitors. *Bioorg. Med. Chem.,* **2012**, *20*(14), 4316-4322.
[http://dx.doi.org/10.1016/j.bmc.2012.05.050] [PMID: 22710102]

[117]   Verma, G.; Chashoo, G.; Ali, A.; Khan, M.F.; Akhtar, W.; Ali, I.; Akhtar, M.; Alam, M.M.; Shaquiquzzaman, M. Synthesis of pyrazole acrylic acid based oxadiazole and amide derivatives as antimalarial and anticancer agents. *Bioorg. Chem.,* **2018**, *77*, 106-124.
[http://dx.doi.org/10.1016/j.bioorg.2018.01.007] [PMID: 29353728]

[118]   Kumar, R.; Kumar, A.; Jain, S.; Kaushik, D. Synthesis, antibacterial evaluation and QSAR studies of 7-[4-(5-aryl-1,3,4-oxadiazole-2-yl)piperazinyl] quinolone derivatives. *Eur. J. Med. Chem.,* **2011**, *46*(9), 3543-3550.
[http://dx.doi.org/10.1016/j.ejmech.2011.04.035] [PMID: 21689870]

[119]   Gordeev, M.F.; Hackbarth, C.; Barbachyn, M.R.; Banitt, L.S.; Gage, J.R.; Luehr, G.W.; Gomez, M.; Trias, J.; Morin, S.E.; Zurenko, G.E.; Parker, C.N.; Evans, J.M.; White, R.J.; Patel, D.V. Novel oxazolidinone-quinolone hybrid antimicrobials. *Bioorg. Med. Chem. Lett.,* **2003**, *13*(23), 4213-4216.
[http://dx.doi.org/10.1016/j.bmcl.2003.07.021] [PMID: 14623004]

[120]   Hubschwerlen, C.; Specklin, J.L.; Baeschlin, D.K.; Borer, Y.; Haefeli, S.; Sigwalt, C.; Schroeder, S.; Locher, H.H. Structure-activity relationship in the oxazolidinone-quinolone hybrid series: influence of the central spacer on the antibacterial activity and the mode of action. *Bioorg. Med. Chem. Lett.,* **2003**, *13*(23), 4229-4233.
[http://dx.doi.org/10.1016/j.bmcl.2003.07.028] [PMID: 14623007]

[121]   Hubschwerlen, C.; Specklin, J.L.; Sigwalt, C.; Schroeder, S.; Locher, H.H. Design, synthesis and biological evaluation of oxazolidinone-quinolone hybrids. *Bioorg. Med. Chem.,* **2003**, *11*(10), 2313-2319.
[http://dx.doi.org/10.1016/S0968-0896(03)00083-X] [PMID: 12713843]

[122]   Wiles, J.A.; Wang, Q.; Lucien, E.; Hashimoto, A.; Song, Y.; Cheng, J.; Marlor, C.W.; Ou, Y.; Podos, S.D.; Thanassi, J.A.; Thoma, C.L.; Deshpande, M.; Pucci, M.J.; Bradbury, B.J. Isothiazoloquinolones containing functionalized aromatic hydrocarbons at the 7-position: synthesis and *in vitro* activity of a series of potent antibacterial agents with diminished cytotoxicity in human cells. *Bioorg. Med. Chem. Lett.,* **2006**, *16*(5), 1272-1276.
[http://dx.doi.org/10.1016/j.bmcl.2005.11.065] [PMID: 16337791]

[123]   Sharma, P.C.; Jain, A.; Yar, M.S.; Pahwa, R.; Singh, J.; Chanalia, P. Novel fluoroquinolone

derivatives bearing N-thiomide linkage with 6-substituted-2-aminobenzothiazoles: Synthesis and antibacterial evaluation. *Arab. J. Chem.,* **2017**, *10*, S568-S575.
[http://dx.doi.org/10.1016/j.arabjc.2012.11.002]

[124] Zhang, H.Z.; Damu, G.L.V.; Cai, G.X.; Zhou, C.H. Design, synthesis and antimicrobial evaluation of novel benzimidazole type of Fluconazole analogues and their synergistic effects with Chloromycin, Norfloxacin and Fluconazole. *Eur. J. Med. Chem.,* **2013**, *64*, 329-344.
[http://dx.doi.org/10.1016/j.ejmech.2013.03.049] [PMID: 23644216]

[125] El-Gohary, N.S.; Shaaban, M.I. Synthesis, antimicrobial, antiquorum-sensing and antitumor activities of new benzimidazole analogs. *Eur. J. Med. Chem.,* **2017**, *137*, 439-449.
[http://dx.doi.org/10.1016/j.ejmech.2017.05.064] [PMID: 28623814]

[126] Grillot, A.L.; Le Tiran, A.; Shannon, D.; Krueger, E.; Liao, Y.; O'Dowd, H.; Tang, Q.; Ronkin, S.; Wang, T.; Waal, N.; Li, P.; Lauffer, D.; Sizensky, E.; Tanoury, J.; Perola, E.; Grossman, T.H.; Doyle, T.; Hanzelka, B.; Jones, S.; Dixit, V.; Ewing, N.; Liao, S.; Boucher, B.; Jacobs, M.; Bennani, Y.; Charifson, P.S. Second-generation antibacterial benzimidazole ureas: discovery of a preclinical candidate with reduced metabolic liability. *J. Med. Chem.,* **2014**, *57*(21), 8792-8816.
[http://dx.doi.org/10.1021/jm500563g] [PMID: 25317480]

[127] Fang, B.; Zhou, C.H.; Rao, X.C. Synthesis and biological activities of novel amine-derived bis-azoles as potential antibacterial and antifungal agents. *Eur. J. Med. Chem.,* **2010**, *45*(9), 4388-4398.
[http://dx.doi.org/10.1016/j.ejmech.2010.06.012] [PMID: 20598399]

[128] Zhang, L.; Addla, D.; Ponmani, J.; Wang, A.; Xie, D.; Wang, Y.N.; Zhang, S.L.; Geng, R.X.; Cai, G.X.; Li, S.; Zhou, C.H. Discovery of membrane active benzimidazole quinolones-based topoisomerase inhibitors as potential DNA-binding antimicrobial agents. *Eur. J. Med. Chem.,* **2016**, *111*, 160-182.
[http://dx.doi.org/10.1016/j.ejmech.2016.01.052] [PMID: 26871658]

[129] Chai, Y.; Liu, M.; Wang, B.; You, X.; Feng, L.; Zhang, Y.; Cao, J.; Guo, H. Synthesis and *in vitro* antibacterial activity of novel fluoroquinolone derivatives containing substituted piperidines. *Bioorg. Med. Chem. Lett.,* **2010**, *20*(17), 5195-5198.
[http://dx.doi.org/10.1016/j.bmcl.2010.07.006] [PMID: 20667731]

[130] Guo, Q.; Feng, L.S.; Liu, M.L.; Zhang, Y.B.; Chai, Y.; Lv, K.; Guo, H.Y.; Han, L.Y. Synthesis and *in vitro* antibacterial activity of fluoroquinolone derivatives containing 3-(N'-alkoxycarbamimidoyl-4-(alkoxyimino) pyrrolidines. *Eur. J. Med. Chem.,* **2010**, *45*(11), 5498-5506.
[http://dx.doi.org/10.1016/j.ejmech.2010.08.050] [PMID: 20833454]

[131] Huang, J.; Liu, H.; Liu, M.; Zhang, R.; Li, L.; Wang, B.; Wang, M.; Wang, C.; Lu, Y. Synthesis, antimycobacterial and antibacterial activity of 1-[(1R,2S)-2-fluorocyclopropyl]naphthyridone derivatives containing an oxime-functionalized pyrrolidine moiety. *Bioorg. Med. Chem. Lett.,* **2015**, *25*(22), 5058-5063.
[http://dx.doi.org/10.1016/j.bmcl.2015.10.027] [PMID: 26476970]

[132] Abdel-Aziz, M.; Park, S.E.; Abuo-Rahma, Gel-D.; Sayed, M.A.; Kwon, Y. Novel *N*-4-piperazin-1-ciprofloxacin-chalcone hybrids: synthesis, physicochemical properties, anticancer and topoisomerase I and II inhibitory activity. *Eur. J. Med. Chem.,* **2013**, *69*, 427-438.
[http://dx.doi.org/10.1016/j.ejmech.2013.08.040] [PMID: 24090914]

[133] Zhang, L.; Kumar, K.V.; Geng, R.X.; Zhou, C.H. Design and biological evaluation of novel quinolone-based metronidazole derivatives as potent Cu^{2+}) mediated DNA-targeting antibacterial agents. *Bioorg. Med. Chem. Lett.,* **2015**, *25*(17), 3699-3705.
[http://dx.doi.org/10.1016/j.bmcl.2015.06.041] [PMID: 26149183]

[134] Cui, S.F.; Peng, L.P.; Zhang, H.Z.; Rasheed, S.; Vijaya Kumar, K.; Zhou, C.H. Novel hybrids of metronidazole and quinolones: synthesis, bioactive evaluation, cytotoxicity, preliminary antimicrobial mechanism and effect of metal ions on their transportation by human serum albumin. *Eur. J. Med. Chem.,* **2014**, *86*, 318-334.
[http://dx.doi.org/10.1016/j.ejmech.2014.08.063] [PMID: 25173851]

[135] Sriram, D.; Bal, T.R.; Yogeeswari, P. Aminopyrimidinimino isatin analogues: design of novel non-nucleoside HIV-1 reverse transcriptase inhibitors with broad-spectrum chemotherapeutic properties. *J. Pharm. Pharm. Sci.,* **2005**, *8*(3), 565-577.
[PMID: 16401403]

[136] Banerjee, D.; Yogeeswari, P.; Bhat, P.; Thomas, A.; Srividya, M.; Sriram, D. Novel isatinyl thiosemicarbazones derivatives as potential molecule to combat HIV-TB co-infection. *Eur. J. Med. Chem.,* **2011**, *46*(1), 106-121.
[http://dx.doi.org/10.1016/j.ejmech.2010.10.020] [PMID: 21093117]

[137] Joy, N.; Mathew, B. Molecular hybridization and preclinical evaluation of imines from para-substituted 4-phenyl 2-amino thiazole incorporated with isatin analogues as antitubercular agents. *Antiinfect. Agents,* **2015**, *13*, 60-64.
[http://dx.doi.org/10.2174/2211352512666140905232639]

[138] Piscopo, E.; Diurno, M.V.; Gagliardi, R.; Cucciniello, M.; Veneruso, G. Studies on heterocyclic compounds: indol-2,3-dione derivatives. VII. Variously substituted hydrazones with antimicrobial activity. *Boll. Soc. Ital. Biol. Sper.,* **1987**, *63*(9), 827-832.
[PMID: 3447623]

[139] Sriram, D.; Bal, T.R.; Yogeeswari, P. Newer aminopyrimidinimino isatin analogues as non-nucleoside HIV-1 reverse transcriptase inhibitors for HIV and other opportunistic infections of AIDS: design, synthesis and biological evaluation. *Farmaco,* **2005**, *60*(5), 377-384.
[http://dx.doi.org/10.1016/j.farmac.2005.03.005] [PMID: 15876436]

[140] Sriram, D.; Yogeeswari, P.; Gopal, G. Synthesis, anti-HIV and antitubercular activities of lamivudine prodrugs. *Eur. J. Med. Chem.,* **2005**, *40*(12), 1373-1376.
[http://dx.doi.org/10.1016/j.ejmech.2005.07.006] [PMID: 16129516]

[141] Xu, Z.; Zhang, S.; Gao, C.; Fan, J.; Zhao, F.; Lv, Z.S.; Feng, L.S. Isatin hybrids and their anti-tuberculosis activity. *Chin. Chem. Lett.,* **2017**, *28*, 159-167.
[http://dx.doi.org/10.1016/j.cclet.2016.07.032]

[142] Panda, S.S.; Detistov, O.S.; Girgis, A.S.; Mohapatra, P.P.; Samir, A.; Katritzky, A.R. Synthesis and molecular modeling of antimicrobial active fluoroquinolone-pyrazine conjugates with amino acid linkers. *Bioorg. Med. Chem. Lett.,* **2016**, *26*(9), 2198-2205.
[http://dx.doi.org/10.1016/j.bmcl.2016.03.062] [PMID: 27025339]

[143] Roth, B.; Rauckman, B.S.; Ferone, R.; Baccanari, D.P.; Champness, J.N.; Hyde, R.M. 2,4-Diamino-5-benzylpyrimidines as antibacterial agents. 7. Analysis of the effect of 3,5-dialkyl substituent size and shape on binding to four different dihydrofolate reductase enzymes. *J. Med. Chem.,* **1987**, *30*(2), 348-356.
[http://dx.doi.org/10.1021/jm00385a017] [PMID: 3100802]

[144] Pokrovskaya, V.; Baasov, T. Dual-acting hybrid antibiotics: a promising strategy to combat bacterial resistance. *Expert Opin. Drug Discov.,* **2010**, *5*(9), 883-902.
[http://dx.doi.org/10.1517/17460441.2010.508069] [PMID: 22823262]

[145] Karoli, T.; Mamidyala, S.K.; Zuegg, J.; Fry, S.R.; Tee, E.H.L.; Bradford, T.A.; Madala, P.K.; Huang, J.X.; Ramu, S.; Butler, M.S.; Cooper, M.A. Structure aided design of chimeric antibiotics. *Bioorg. Med. Chem. Lett.,* **2012**, *22*(7), 2428-2433.
[http://dx.doi.org/10.1016/j.bmcl.2012.02.019] [PMID: 22406152]

[146] Huovinen, P.; Wolfson, J.S.; Hooper, D.C. Synergism of trimethoprim and ciprofloxacin *in vitro* against clinical bacterial isolates. *Eur. J. Clin. Microbiol. Infect. Dis.,* **1992**, *11*(3), 255-257.
[http://dx.doi.org/10.1007/BF02098092] [PMID: 1597204]

[147] Perez-Vizcaino, F.; Duarte, J.; Santos-Buelga, C. The flavonoid paradox: conjugation and deconjugation as key steps for the biological activity of flavonoids. *J. Sci. Food Agric.,* **2012**, *92*(9), 1822-1825.
[http://dx.doi.org/10.1002/jsfa.5697] [PMID: 22555950]

[148]   Hosseinimehr, S.J. Flavonoids and genomic instability induced by ionizing radiation. *Drug Discov. Today,* **2010**, *15*(21-22), 907-918.
[http://dx.doi.org/10.1016/j.drudis.2010.09.005] [PMID: 20933097]

[149]   Wong, I.L.K.; Chan, K.F.; Burkett, B.A.; Zhao, Y.; Chai, Y.; Sun, H.; Chan, T.H.; Chow, L.M. Flavonoid dimers as bivalent modulators for pentamidine and sodium stiboglucanate resistance in leishmania. *Antimicrob. Agents Chemother.,* **2007**, *51*(3), 930-940.
[http://dx.doi.org/10.1128/AAC.00998-06] [PMID: 17194831]

[150]   Wong, I.L.K.; Chan, K.F.; Tsang, K.H.; Lam, C.Y.; Zhao, Y.; Chan, T.H.; Chow, L.M.C. Modulation of multidrug resistance protein 1 (MRP1/ABCC1)-mediated multidrug resistance by bivalent apigenin homodimers and their derivatives. *J. Med. Chem.,* **2009**, *52*(17), 5311-5322.
[http://dx.doi.org/10.1021/jm900194w] [PMID: 19725578]

[151]   Jaganathan, S.K. Can flavonoids from honey alter multidrug resistance? *Med. Hypotheses,* **2011**, *76*(4), 535-537.
[http://dx.doi.org/10.1016/j.mehy.2010.12.011] [PMID: 21247706]

[152]   Aboul-Fadl, T.; Mohammed, F.A.H.; Hassan, E.A.S. Synthesis, antitubercular activity and pharmacokinetic studies of some Schiff bases derived from 1-alkylisatin and isonicotinic acid hydrazide (INH). *Arch. Pharm. Res.,* **2003**, *26*(10), 778-784.
[http://dx.doi.org/10.1007/BF02980020] [PMID: 14609123]

[153]   Sridhar, S.K.; Saravanan, M.; Ramesh, A. Synthesis and antibacterial screening of hydrazones, Schiff and Mannich bases of isatin derivatives. *Eur. J. Med. Chem.,* **2001**, *36*(7-8), 615-625.
[http://dx.doi.org/10.1016/S0223-5234(01)01255-7] [PMID: 11600231]

[154]   Bame, J.R.; Graf, T.N.; Junio, H.A.; Bussey, R.O., III; Jarmusch, S.A.; El-Elimat, T.; Falkinham, J.O., III; Oberlies, N.H.; Cech, R.A.; Cech, N.B. Sarothrin from Alkanna orientalis is an antimicrobial agent and efflux pump inhibitor. *Planta Med.,* **2013**, *79*(5), 327-329.
[http://dx.doi.org/10.1055/s-0032-1328259] [PMID: 23468310]

[155]   Kerns, R.J.; Rybak, M.J.; Kaatz, G.W.; Vaka, F.; Cha, R.; Grucz, R.G.; Diwadkar, V.U. Structural features of piperazinyl-linked ciprofloxacin dimers required for activity against drug-resistant strains of Staphylococcus aureus. *Bioorg. Med. Chem. Lett.,* **2003**, *13*(13), 2109-2112.
[http://dx.doi.org/10.1016/S0960-894X(03)00376-7] [PMID: 12798315]

[156]   Ross, A.G.; Benton, B.M.; Chin, D.; De Pascale, G.; Fuller, J.; Leeds, J.A.; Reck, F.; Richie, D.L.; Vo, J.; LaMarche, M.J. Synthesis of ciprofloxacin dimers for evaluation of bacterial permeability in atypical chemical space. *Bioorg. Med. Chem. Lett.,* **2015**, *25*(17), 3468-3475.
[http://dx.doi.org/10.1016/j.bmcl.2015.07.010] [PMID: 26189081]

[157]   Khan, M.; Reddy, C.N.; Ravindra, G.; Reddy, K.V.S.R.K.; Dubey, P.K. Development and validation of a stability indicating HPLC method for simultaneous determination of four novel fluoroquinolone dimers as potential antibacterial agents. *J. Pharm. Biomed. Anal.,* **2012**, *59*, 162-166.
[http://dx.doi.org/10.1016/j.jpba.2011.09.025] [PMID: 22030076]

[158]   Hu, G.Q.; Zhang, Z.Q.; Xie, S.Q.; Huang, W.L. Synthesis and antitumor evaluation of C3/C3 fluoroquinolone dimers (I): Tethered with a fused heterocyclic s-triazolo[2,1-b][1,3,4]thiadiazole. *Chin. Chem. Lett.,* **2010**, *21*, 661-663.
[http://dx.doi.org/10.1016/j.cclet.2010.01.037]

[159]   Detsi, A.; Bouloumbasi, D.; Prousis, K.C.; Koufaki, M.; Athanasellis, G.; Melagraki, G.; Afantitis, A.; Igglessi-Markopoulou, O.; Kontogiorgis, C.; Hadjipavlou-Litina, D.J. Design and synthesis of novel quinolinone-3-aminoamides and their alpha-lipoic acid adducts as antioxidant and anti-inflammatory agents. *J. Med. Chem.,* **2007**, *50*(10), 2450-2458.
[http://dx.doi.org/10.1021/jm061173n] [PMID: 17444626]

[160]   Teichert, A.; Schmidt, J.; Porzel, A.; Arnold, N.; Wessjohann, L. (Iso)-quinoline alkaloids from fungal fruiting bodies of Cortinarius subtortus. *J. Nat. Prod.,* **2008**, *71*(6), 1092-1094.
[http://dx.doi.org/10.1021/np8000859] [PMID: 18447384]

[161]   Carlier, P.R.; Du, D-M.; Han, Y-F.; Liu, J.; Perola, E.; Williams, I.D.; Pang, Y.P. Dimerization of an inactive fragment of huperzine A produces a drug with twice the potency of the natural product. *Angew. Chem. Int. Ed. Engl.,* **2000**, *39*(10), 1775-1777.
[http://dx.doi.org/10.1002/(SICI)1521-3773(20000515)39:10<1775::AID-ANIE1775>3.0.CO;2-Q] [PMID: 10934357]

[162]   Pudlo, M.; Luzet, V.; Ismaïli, L.; Tomassoli, I.; Iutzeler, A.; Refouvelet, B. Quinolone-benzylpiperidine derivatives as novel acetylcholinesterase inhibitor and antioxidant hybrids for Alzheimer disease. *Bioorg. Med. Chem.,* **2014**, *22*(8), 2496-2507.
[http://dx.doi.org/10.1016/j.bmc.2014.02.046] [PMID: 24657052]

[163]   León, R.; Garcia, A.G.; Marco-Contelles, J. Recent advances in the multitarget-directed ligands approach for the treatment of Alzheimer's disease. *Med. Res. Rev.,* **2013**, *33*(1), 139-189.
[http://dx.doi.org/10.1002/med.20248] [PMID: 21793014]

[164]   Cavalli, A.; Bolognesi, M.L.; Minarini, A.; Rosini, M.; Tumiatti, V.; Recanatini, M.; Melchiorre, C. Multi-target-directed ligands to combat neurodegenerative diseases. *J. Med. Chem.,* **2008**, *51*(3), 347-372.
[http://dx.doi.org/10.1021/jm7009364] [PMID: 18181565]

[165]   Liu, Q.; Xie, F.; Rolston, R.; Moreira, P.I.; Nunomura, A.; Zhu, X.; Smith, M.A.; Perry, G. Prevention and treatment of Alzheimer disease and aging: antioxidants. *Mini Rev. Med. Chem.,* **2007**, *7*(2), 171-180.
[http://dx.doi.org/10.2174/138955707779802552] [PMID: 17305591]

[166]   Robertson, G.T.; Bonventre, E.J.; Doyle, T.B.; Du, Q.; Duncan, L.; Morris, T.W.; Roche, E.D.; Yan, D.; Lynch, A.S. *In vitro* evaluation of CBR-2092, a novel rifamycin-quinolone hybrid antibiotic: studies of the mode of action in Staphylococcus aureus. *Antimicrob. Agents Chemother.,* **2008**, *52*(7), 2313-2323.
[http://dx.doi.org/10.1128/AAC.01649-07] [PMID: 18443108]

[167]   Robertson, G.T.; Bonventre, E.J.; Doyle, T.B.; Du, Q.; Duncan, L.; Morris, T.W.; Roche, E.D.; Yan, D.; Lynch, A.S. *In vitro* evaluation of CBR-2092, a novel rifamycin-quinolone hybrid antibiotic: microbiology profiling studies with staphylococci and streptococci. *Antimicrob. Agents Chemother.,* **2008**, *52*(7), 2324-2334.
[http://dx.doi.org/10.1128/AAC.01651-07] [PMID: 18443106]

[168]   Ma, Z.; Lynch, A.S. Development of a Dual-Acting Antibacterial Agent (TNP-2092) for the Treatment of Persistent Bacterial Infections. *J. Med. Chem.,* **2016**, *59*(14), 6645-6657.
[http://dx.doi.org/10.1021/acs.jmedchem.6b00485] [PMID: 27336583]

[169]   Emami, S.; Shafiee, A.; Foroumadi, A. Structural features of new quinolones and relationship to antibacterial activity against Gram-positive bacteria. *Mini Rev. Med. Chem.,* **2006**, *6*(4), 375-386.
[http://dx.doi.org/10.2174/138955706776361493] [PMID: 16613574]

[170]   Pokrovskaya, V.; Belakhov, V.; Hainrichson, M.; Yaron, S.; Baasov, T. Design, synthesis, and evaluation of novel fluoroquinolone-aminoglycoside hybrid antibiotics. *J. Med. Chem.,* **2009**, *52*(8), 2243-2254.
[http://dx.doi.org/10.1021/jm900028n] [PMID: 19301822]

CHAPTER 3

Structure of Fine Starch Prepared *Via* a Compressed Hot Water Process

N. Shimizu[1,2,*] and **T. Ushiyama**[3]

[1] *Research Faculty of Agriculture, Hokkaido University, Hokkaido, 060-8589, Japan*

[2] *Field Science Center for Northern Biosphere, Hokkaido University, 060-0811, Japan*

[3] *Graduate School of Agriculture, Hokkaido University, Hokkaido, 060-8589, Japan*

Abstract: In a "top-down" process, starch nanoparticles can be produced by structural and size refinements through the breakdown of large particles. In this study, the structure of fine starches prepared *via* a compressed hot water process at different temperatures (160 – 180°C) was analysed using dynamic light scattering and size-exclusion chromatography with multi-angle light scattering (MALS) and differential refractive index detection. Changes in the molecular weight, polydispersity, hydrodynamic radius, and radius of gyration were assessed. The intrinsic viscosity of the fine starch solution was derived from the Flory-Fox and Ptitsyn-Eizner equation. The weight-average molecular weight decreased to 7.29×10^{-5} g/mol while the average hydrodynamic radius and weight-average radius of gyration decreased by 34.9 nm and 14.6 nm respectively, in fine starch prepared at 180 °C. In fine starches prepared at 160 °C, 165 °C, and 170 °C, tails in the multi-angle light scattering peaks, upswings in the conformation plots, and upturns in the plots of gyration radii and elution volumes were all the result of branching structures. In fine starches prepared at 175 °C and 180 °C, amylopectin branching was diminished and symmetrical scattering peaks were detected in the MALS analysis. We propose a pathway for waxy rice starch hydrolysis by a compressed hot water process.

Keywords: Fine starch, Intrinsic viscosity, Hydrodynamic radius, Molecular weight, Radius of gyration.

INTRODUCTION

Starch Conformation in Solid State and in Solution

Starch is a renewable and biodegradable biopolymer that is stored in many plants as a source of energy for photosynthesis. It is the second most abundant biomass in nature, and is typically isolated from plants in the form of microscale granules.

* **Corresponding author Shimizu N:** Research Faculty of Agriculture and Field Science Center for Northern Biosphere, Hokkaido University, Japan; Tel/Fax: +81-11-706-3848; E-mail: shimizu@bpe.agr.hokudai.ac.jp

Atta-ur-Rahman (Ed.)

Fine starches with average particle sizes ranging from the micrometre to nanometre scale have been developed as functional materials with applications in foods, cosmetics, medicines, and various composites. The major characteristics of fine starches are rapid dissolution and enhanced bioavailability after consumption.

Recent studies have reported that nano-scale starch particles can be readily prepared from starch granules, which have unique physical properties [1]. Starch granules consist of numerous nano-size semi-crystalline blocklets [2]. Physical treatments may disintegrate the starch granules, thus releasing the nano-blocklets. The preparation of starch nanoparticles may be classified into "top-down" and "bottom-up" processes. In the "top-down" process, nanoparticles are produced from structural and size refinement through the breakdown of large particles [3, 4]. In the "bottom-up" process, starch nanoparticles self-assemble into starch particles. Starch nanoparticles are important vehicles for nano- and micro-encapsulation in the food industry [5 - 7]. The physicochemical properties of polymers such as molecular weight, polydispersity, radius of gyration, hydrodynamic radius, and the molecular structure in solution are derived before and after modification processes [8, 9]. The intrinsic viscosity of the fine starch solution can be derived using the Flory-Fox and Ptitsyn-Einzer equation. It is important to evaluate not only raw starch, but also modified starches, for industrial use [3, 4].

To use starch effectively, techniques to prepare nanoscale starches have been developed and assessed. Nano-scale waxy rice starch particles can be prepared *via* hydrolysis using a compressed hot water process. Starch nanoparticles can also be prepared by acid hydrolysis, enzymatic treatments, and physical treatments such as high-pressure homogenization, ultrasonication, reactive extrusion, and gamma irradiation [1]. The compressed hot water process is one of the most useful reactions. The smallest average hydrodynamic radius of 75.2 nm was obtained by using a 180°C compressed hot water treatment, with a starch concentration of 0.1% (w/w), and an initial pressure of 3.0 MPa. The product of this process was evaluated by zeta potential and by using a submicron particle size analyzer [10]. A fine waxy rice starch solution prepared at 160°C with a compressed hot water treatment was spray-dried as a wall material for micro encapsulation [6]. It is important to understand starch macromolecular structures to optimize the practical uses of various products in industry. However, the changes in the hydrodynamic particle size induced by the compressed hot water process are unknown.

In this chapter, we describe the structure of fine starches prepared *via* a compressed hot water process at different temperatures (160°C – 180°C). The fine starches were analysed using dynamic light scattering (DLS) and size-exclusion chromatography (SEC) with multi-angle light scattering (MALS) and differential refractive index (DRI) detection.

The intrinsic viscosity was calculated using the Flory-Fox and Ptitsyn-Eizner equation and correlations with particle conformation were derived. In SEC-MALS measurements, the branching amylopectin polymers affected the column separation and the MALS signal. Therefore, changes in the branching structures in waxy rice starch by the compressed hot water process could be deduced from the SEC-MALS data.

Starch Composition and Chemical Structure

Starch is commonly extracted from corn, wheat, and tapioca in native and modified forms for applications in the food, paper, and pharmaceutical industries [11 - 14]. After extraction from plants, starch occurs as flour-like white particles that are insoluble in cold water. For example, rice starch granules are semi-crystalline particles ranging from 3 to 8 μm. The smallest known starch particles are those in cereal grains. There is some variation in starch granule size among different rice genotypes. Rice starch granules have a smooth surface, but angular and polygonal shapes. The granules are loosely packed in clusters, and some particles have holes and cracks (Fig. **1**).

The internal architecture of native starch granules is characterized by "growth rings" that represent concentric semi-crystalline shells (thickness 120 – 400 nm) separated by amorphous regions. There is evidence that the crystalline shells consist of regular alternating amorphous and crystalline lamellae repeating at 9 – 10 nm intervals. In this structural organization, parallel double helices of amylopectin side chains assemble into radially oriented clusters (Fig. **2**). Little is known about the structure, organization, and arrangement of the lamellae.

Starch consists of amylase α(1-4)-linked glucose units, amylopectin α(1-4)-linked glucose units, and branched α(1-6)-linkages. The molecular weights (M_w) of amylose and amylopectin in starch vary among different plants; in normal corn they are 1.4×10^6 and 39×10^6 g/mol, respectively [15]; that of amylose from rice is $5.1\text{-}6.9 \times 10^5$ g/mol [16]; that of amylose from waxy barley starch is 1.06×10^8 g/mol [17]; and that of amylose in Amioca (waxy corn starch) ranges from $107\text{-}10^9$ g/mol [18].

Starch molecules have a semi-crystalline structure that significantly affects their physical and chemical performance. Analyses of the fine waxy starch after the compressed hot water process revealed a peak derived from its crystalline structure. This peak was much reduced after the hydrothermal and spray-drying processes, indicating that the crystalline structure in the starch molecules had broken down into the amorphous form.

Fig. (1). Scanning electron microscope image of waxy rice starch granules.

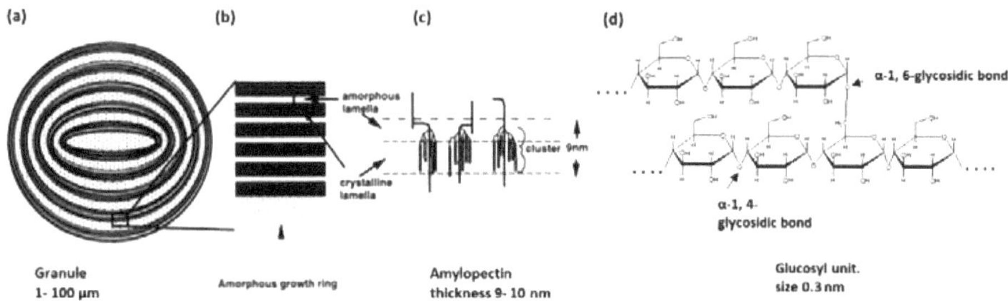

Fig. (2). Schematic representation of the different structural levels of the starch granule and the involvement of amylopectin.

Preparation of Starch Nanoparticles

In the compressed hot water process, fine starch is produced from biomass using only hot water under high pressure, without the need for catalysts, acids, or enzymes. This process is one of the most useful reactions, as shown in the phase diagram (Fig. **3**). The physical state of the compressed hot water is controlled by temperature and pressure, and is reflected by the specific dielectric constant, ranging from 40 to 60, ion product ranging from −11.4 to −12.3, and the internal energy ranging from 7 to 15 kJ/mol (Fig. (**4a & 4b**).

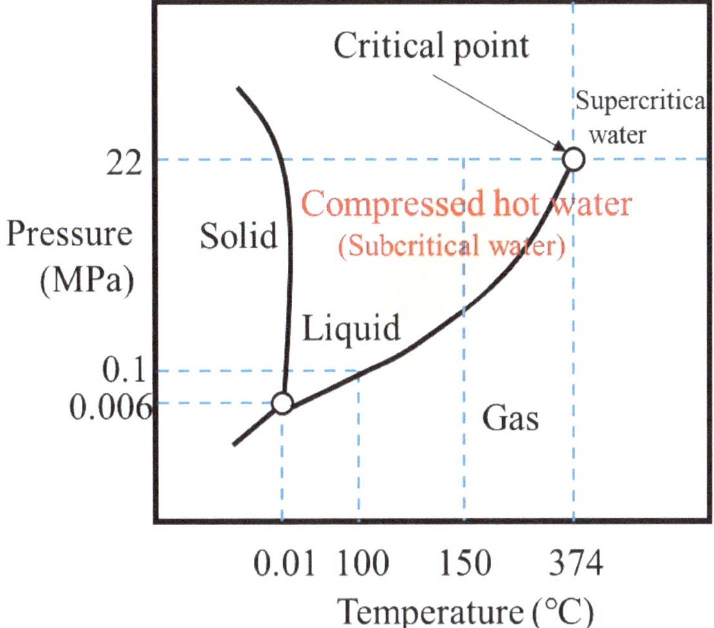

Fig. (3). State diagram of water.

A pressure vessel (stainless-steel vessel: φ48 mm, Teflon resin inner vessel: φ36 mm × 120 mm tall) was used for the compressed hot water process (Fig. **4a**). To prepare fine starch solutions by the compressed hot water process, 50 ml distilled water was added to 0.05 g starch in the pressure vessel of an organic synthesizer, stirred with a stirring chip, and heated until the temperature of the solution was 160 °C, 165 °C, 170 °C, 175 °C, or 180 °C, then cooled in a constant temperature bath at 5 °C. All solutions were then stored in a refrigerator at 5 °C until analysis (Fig. **5**).

SEC-MALS, DLS, and Intrinsic Viscosity Analyses

Size-exclusion chromatography with multi-angle light scattering (SEC-MALS) and field-flow fractionation with multi-angle light scattering (FFF-MALS) are powerful methods for the fractionation and characterization of macromolecules. The physicochemical properties of polymers such as the molecular weight, polydispersity, radius of gyration, hydrodynamic radius, and the molecular structure in a solution were derived using these methods [8, 9]. The intrinsic viscosity of each fine starch solution was derived from the Flory-Fox and Ptitsyn-Einzer equation. Evaluation of not only raw starch but also modified starches is important for industrial use [3, 4].

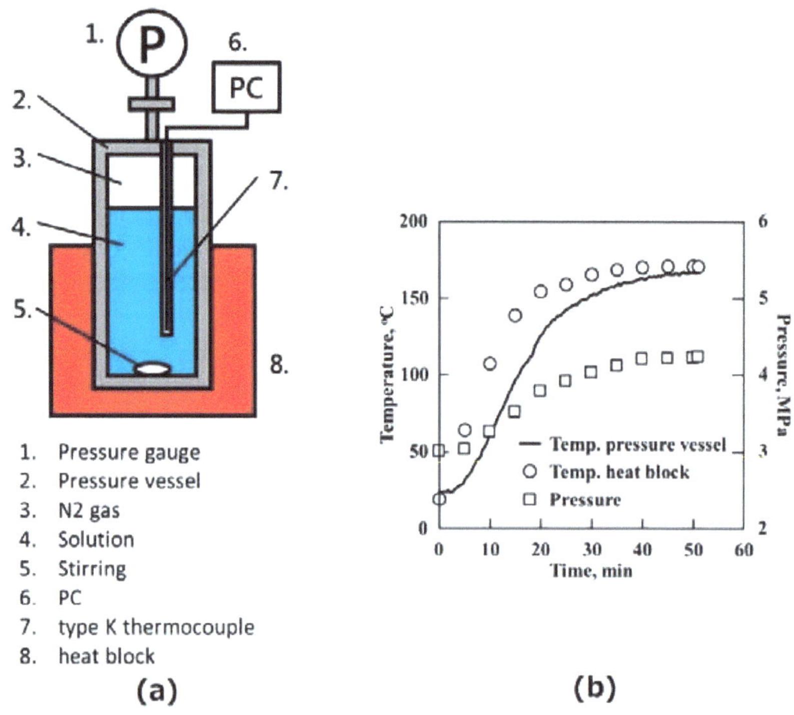

Fig. (4). Schematic of a compressed hot water apparatus (**a**) and change in temperature and pressure during its process (**b**).

SEC-MALS Analysis

The instruments used for SEC-MALS were a high-performance liquid chromatograph (HPLC 1200 Infinity series, Agilent Technologies, Santa Clara, CA), a multi-angle light scattering photometer (MALS DAWN8+, Wyatt Technology, Santa Barbara, CA), and a differential refractive index (RI) detector (OptilabrEX, Wyatt Technology, Santa Barbara, CA, USA). For MALS and RI, a wavelength of 658 nm was used. The HPLC (Showa Denko K.K., Tokyo, Japan) was equipped with a liquid delivery pump, an automatic sampler, a column compartment, and a UV detector. A LB-G6B guard column and a LB-806M analytical size-exclusion column were connected in tandem, and were kept within the column compartment at 30°C. The buffer solution for separations was 50-mM NaNO$_3$ in Milli-Q water that was passed through a 0.45-µm filter (MF-Millipore membrane filters, Merck KGaA, Germany) before use. The 50-µL sample injection was started at a flow velocity of 0.8 mL/min after the laser intensity for detection in MALS and RI stabilized. For detector signal alignment, we used a 3 mg/mL pullulan standard solution (PSS-dpul50k, PSS polymer Standard Service GmbH, Mainz, Germany).

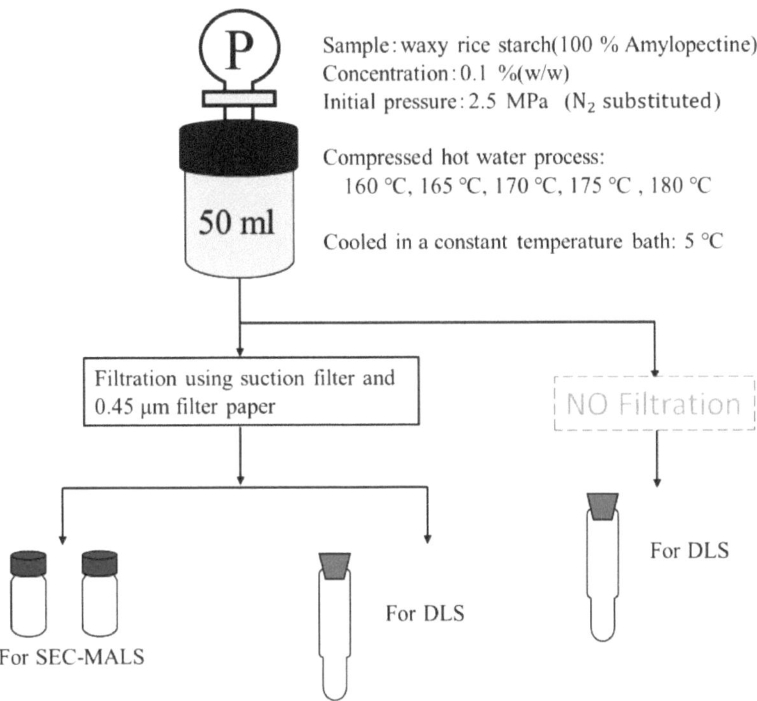

Fig. (5). Preparation of starch nanoparticles.

The SEC-MALS data were processed using Astra software 5.3.4 (Wyatt Technology, Santa Barbara, California) which used the upper 50% of the light scattering peak intensity. The collection interval in MALS was 0.5 s, and each data slice was equivalent to 0.67 µL. The refractive index increment (dn/dc) of 0.146 mL/g was used for fine starches prepared by the compressed hot water process [16, 17, 19]. The molecular weight and the radius of gyration were determined by Zimm plots [20]. The data from the MALS detector were fitted to a straight line on the Zimm plots. Inaccurate data, especially those obtained from the low scattering angle detector, were not used. The second coefficient was set to zero because of the low concentration of starch in the samples.

The following information were calculated by ASTRA 5.3.4 by integration over one peak: number-average molecular weight M_n; weight-average molecular weight M_w; Z-average molecular weight M_z; number - the average radius of gyration $\langle R_g \rangle_n$; weight-average radius of gyration $\langle R_g \rangle_w$; and Z-average radius of gyration $\langle R_g \rangle_z$.

$$M_n = \sum \frac{c_i}{\sum(c_i/M_i)} \tag{1}$$

$$M_w = \sum (c_i M_i) / \sum c_i \tag{2}$$

$$M_z = \sum (c_i M_i^2) / \sum (c_i M_i) \tag{3}$$

$$\langle R_g \rangle_n^2 = \sum (c_i \langle R_g \rangle_i^2 / M_i) / \sum (c_i / M_i) \tag{4}$$

$$\langle R_g \rangle_w^2 = \sum (c_i \langle R_g \rangle_i^2) / \sum c_i \tag{5}$$

$$\langle R_g \rangle_z^2 = \sum (c_i M_i \langle R_g \rangle_i^2) / \sum (c_i M_i) \tag{6}$$

where C_i, M_i and $\langle R_g \rangle_i$ are the mass concentration, molecular weight, and radius of gyration of the ith slice, respectively.

The two polydispersity values M_w/M_n and M_z/M_n were calculated. The molecular shape information was obtained from the conformation plot, which is a plot of the logarithm of the radius of gyration *vs.* the logarithm of the molecular weight.

$$\langle R_g \rangle = kM^v \tag{7}$$

The value of the exponent v was obtained from the slope of $\log \langle R_g \rangle$ *vs.* $\log M$.

The molecular structure was also estimated from this slope.

The V-values of 0.33, 1.00, and 0.50 – 0.60 were estimated for a spherical structure, rod structure, and random coil structure, respectively [8].

DLS Analysis

The hydrodynamic radius R_h was obtained by fiber dynamic light scattering analysis (FDLS-3000, Otsuka Electronics Co., Ltd., Japan). Each sample solution was kept in a 12 mm-diameter glass cell (Otsuka Electronics Co., Ltd., Japan) that had been cleaned with chloroform. The measuring device was filled with silicone oil and thermostatically controlled at 25°C. A 532 nm laser was used and the measurement angle was 90°.

R_h was calculated from the Stokes-Einstein equation, as follows:

$$R_h = k_B T / 6\pi\eta D_i \tag{8}$$

where K_B, T, η, and D_i are Boltzmann's constant, absolute temperature, solvent viscosity, and the diffusion coefficient, respectively [21]. The sum of 300 particles was calculated. R_h distributions and the average particle size were obtained with the CONTIN routine.

To evaluate the dispersion of particle size distribution, the relative span factor *RSF* was derived as follows:

$$RSF = (D_{90} - D_{10})/D_{50} \tag{9}$$

Where D_n is the hydrodynamic radius at n% in the cumulative frequency distribution. Small *RSF* values indicate uniform particle distribution [22].

The hydrodynamic radius of distribution of fine starch was expected to change because of passage through the 0.45-μm filter. Therefore, R_h measurements were obtained for samples pre- and post-filtering (Fig. **6**).

Derivation of Intrinsic Viscosity

The intrinsic viscosity [η] is defined as follows:

$$[\eta] = \lim_{c \to 0} (\eta - \eta_0)/\eta_0 c \tag{10}$$

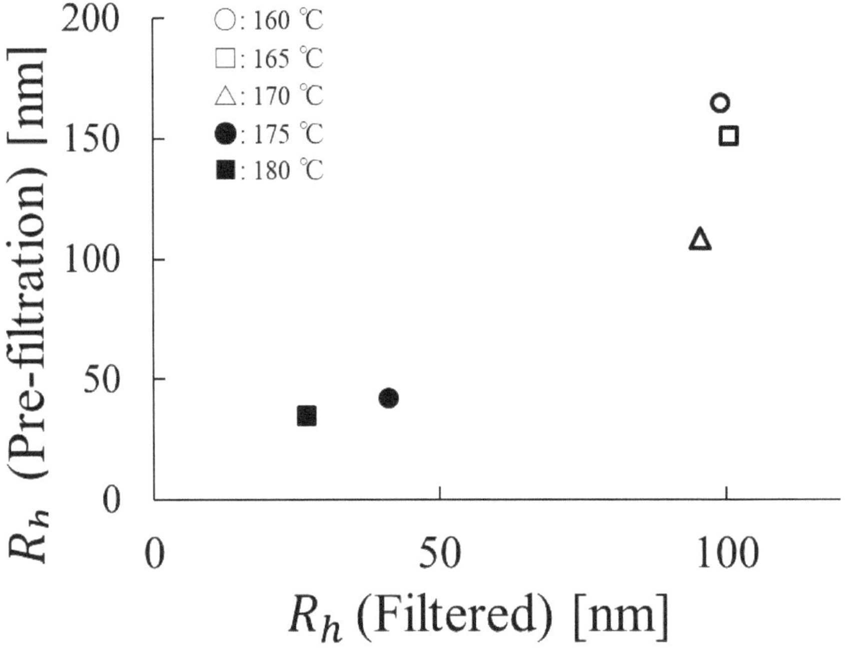

Fig. (6). Hydrodynamic radius R_h obtained from DLS with different compressed hot water process temperature at 160 °C (○), 165 °C (□), 170 °C (□), 175 °C (●) and 180 °C (■) [26].

Where η and η_0 are the viscosities of the dilute polymer solution and the pure solvent, respectively, and c is the polymer concentration. Additionally, $[\eta]$ was calculated using the Flory-Fox and Ptitsyn-Eizner equation [23, 24], which was applied under the non-theta condition solutions:

$$\langle R_g \rangle = \frac{1}{\sqrt{6}} \left(\frac{[\eta]M}{\Phi} \right)^{\frac{1}{3}} \tag{11}$$

$$\Phi = 2.86 \times 10^{21}(1 - 2.63\varepsilon + 2.86\varepsilon^2) \tag{12}$$

$$a = 3v - 1 \tag{13}$$

$[\eta]$, $\langle R \rangle$, and M are the intrinsic viscosity (100 mL/g), the radius of gyration (cm), and the molecular weight (g/mol), respectively. Φ is the corrected Flory universal constant for non-theta conditions and $\Phi 0 = 2.86 \times 10^{21}$ is applied for theta conditions ($\varepsilon = 0$). The value a is the exponent of the Mark-Houwink equation [16], which is expressed as the relationship between $[\eta]$ and M:

$$[\eta] = KM^a \tag{14}$$

Eq. 15 indicates the relationship between a and the slope of conformation plot v:

$$a = 3v - 1 \tag{15}$$

Particle conformation data from SEC-MALS

A single peak was confirmed in the MALS signal for all the samples. The tail at high volumes indicated that the MALS signal was detected after the peak, and this tail was detected in analyses of the fine starches prepared at 160 °C, 165 °C, and 170 °C (Fig. **7A**). Fine starches prepared at 175°C and 180 °C exhibited a symmetrical peak in the MALS analysis because the reduced branching structures in the amylopectin and the decreased particle sizes *via* hydrolysis reduced the tails (Fig. **7B**). In this study, the radius of gyration $\langle R_g \rangle$ and molecular weight (M) decreased with increasing processing temperature. The proportion of branching linkages in the waxy rice starch ranged between 4.0% and 5.5% [25]. The tails occurred because the fine starch particles with branched amylopectin structures, large particles, and particle aggregates were caught in the column filter [9]. Table **1** shows the results obtained from SEC-MALS, including the average molecular weight *(M)*, polydispersity, the average radius of gyration (R_g), slope of the conformation plots (v), and intrinsic viscosity (η) The weight-average molecular weight (M_w) and weight-average radius of gyration ($\langle R_g \rangle$) decreased by 0.729×10^6g/mol and 14.6 nm, respectively, in starch prepared at 180 °C. In general, amylopectin has a high molecular weight, up to 1×10^9g/mol [11]. The compressed hot water process is one of the most powerful hydrolysis reaction fields for polymers; in this study, the molecular weight of amylopectin decreased by three to four orders of magnitude [26]. Fine particles are in the $\langle R_g \rangle_w$ of 36.4 to 68.2 nm, nano particles are in the $\langle R_g \rangle_w$ of 14.6 nm, the macromolecular structure describes the arrangement of polymer chain within the particles. (Fig. **8**) plots the relationship between M_w and the particle radius (average R_h) after filtration and $\langle R_g \rangle_w$. These correlations were strong, where the coefficient of determination R^2 between M_w and R_h was 0.96, and that for $\langle R_g \rangle_w$ was 0.98. The molecular weight and particle size may be inferred from this approximation.

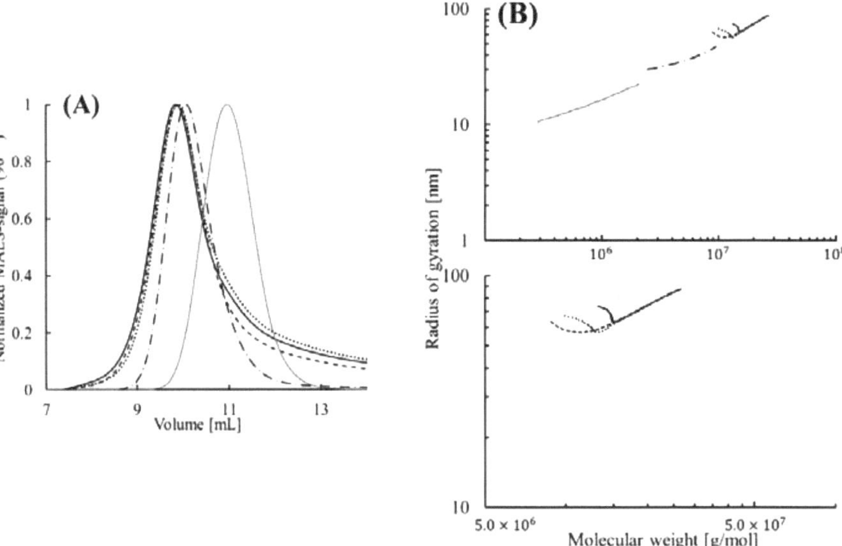

Fig. (7). SEC-MALS results for fine starch prepared by the compressed hot water process at 160 °C (____), 165 °C (.........), 170 °C (_ _ _), 175 °C (_ _ _), and 180 °C (____). (**A**): volume *vs.* detector signals obtained from MALS photometer at 90 °, (**B**): conformation plot and enlarged conformation plot for 160 °C, 165 °C, and 170 °C [26].

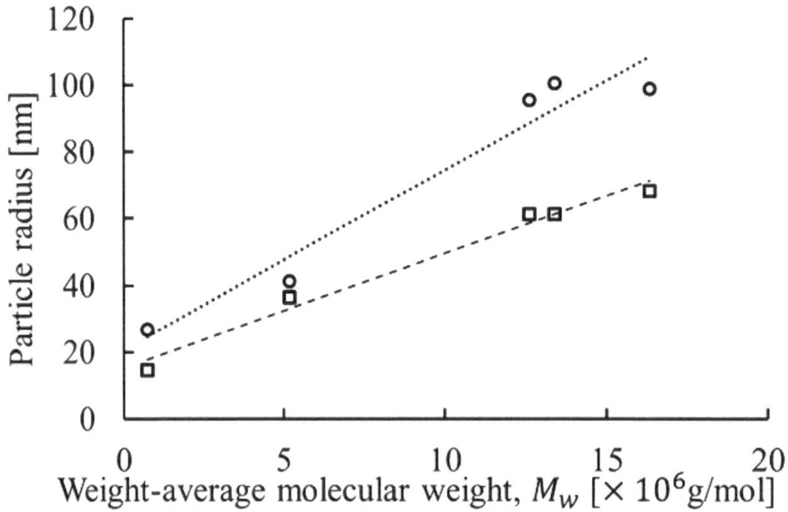

Fig. (8). Relationship between weight-average molecular weight and particle radius. (○): Average R_h after filtration (□): weight-average radius of gyration, $\langle R_g \rangle_w$. The coefficient of determination R^2 was (○) 0.96, (□) 0.98 by linear approximation [26].

Table 1. Results obtained from SEC-MALS [26].

Compressed Hot Water Process	M_n [x10⁶ g/mol]	M_w [x10⁶ g/mol]	M_z [x10⁶ g/mol]	Polydispersity (M_w/M_n)	Polydispersity (M_z/M_n)	$<Rg>n$	$<Rg>w$	$<Rg>z$	Slop Conformation Plot	Intrinsic Viscosity [100mL/g]
160°C	16.0	16.3	16.7	1.02	1.02	67.7	68.2	68.9	0.40	1.34
165°C	13.1	13.4	13.7	1.02	1.02	61.2	61.3	61.6	0.38	1.30
170°C	12.0	12.6	13.3	1.05	1.05	60.5	61.2	62.2	0.32	1.77
175°C	4.49	5.18	5.89	1.15	1.15	34.8	36.4	38.1	0.35	0.801
180°C	0.562	0.729	0.954	1.30	1.30	13.2	14.6	16.2	0.40	0.294

A pathway for waxy rice starch hydrolysis by compressed hot water is presented in (Scheme. S1). It shows natural amylopectin as a randomly branched structure with α(1-6) linkages and 4.0% to 5.5% branching linkages. It further shows that the branching and long chains that affected column fractionation in SEC were gradually degraded by the compressed hot water treatment. Precise column fractionation was obtained through intermolecular dehydration by amylopectin hydrolysis at 175 °C and 180 °C. Starch is the second most abundant material in biomass and is already widely used in industry, so it is widely available and inexpensive. Fine starches are new materials with potential applications in nano- and micro-encapsulation by spray-drying [6].

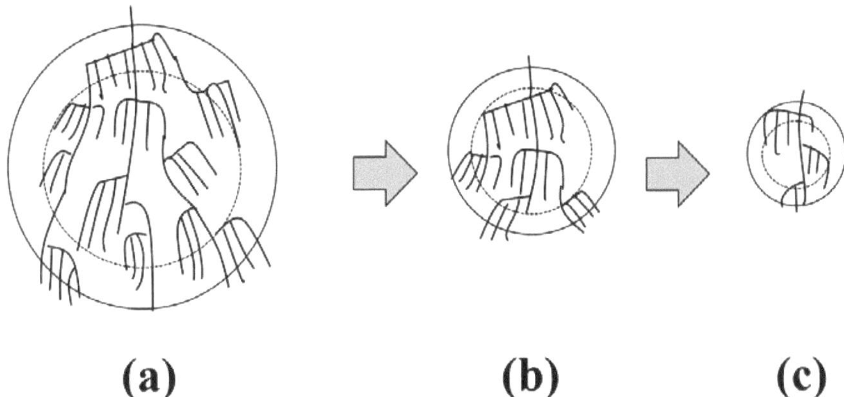

Scheme (1). Waxy rice starch hydrolysis pathway in the compressed hot water process. (a) fine starch prepared at 160 °C with an average R_h=98.9 nm, and $\langle R_g \rangle_w$=68.2 nm, (b) fine starch at 175 °C with an average R_h=41.2 nm, and $\langle R_g \rangle_w$=36.4 nm, (c) nano particle starch at 180°C with an average Rh=26.8 nm, and $\langle R_g \rangle_w$=14.6 nm. The solid-line and broken-line circles indicate hydrodynamic particle size and radius of gyration [26].

CONCLUSION

The macromolecular structure of fine starch prepared by a compressed hot water process was analysed with DLS and SEC-MALS. The branching structure and the long chains of amylopectin were diminished by the compressed hot water process *via* hydrolysis. The intrinsic viscosity was derived from the Flory-Fox and Ptitsyn-Eizner equation and it correlated with the molecular weight, the hydrodynamic radius, and the radius of gyration. In fine starches prepared at 175 °C and 180 °C, amylopectin branching was diminished and symmetric scattering peaks were observed in the MALS analysis. Finally, the weight-average molecular weight M_w and the weight-average radius of gyration $\langle R_g \rangle_w$ were decreased by 0.729×10^6 g/mol and 14.6 nm, respectively, in fine starch prepared at 180 °C. The average hydrodynamic radius was decreased by 34.9 nm in the DLS measurement. We propose a pathway for waxy rice starch hydrolysis by a compressed hot water process. The starch nanoparticles prepared *via* compressed hot water processes at 175 °C and 180 °C were well fitted to methods for the determination of molecular structure.

CONSENT FOR PUBLICATION

Not applicable.

CONFLICT OF INTEREST

The authors confirm that this chapter contents have no conflict of interest.

ACKNOWLEDGEMENTS

Fiber dynamic light scattering and SEC-MALS were provided by the instrumental services of the Global Facility Center of Hokkaido University.

REFERENCES

[1]　Kim, H.Y.; Park, S.S.; Lim, S.T. Preparation, characterization and utilization of starch nanoparticles. *Colloids Surf. B Biointerfaces,* **2015,** *126,* 607-620.
[http://dx.doi.org/10.1016/j.colsurfb.2014.11.011] [PMID: 25435170]

[2]　Gallant, D.J.; Bouchet, P.M.; Baldwin, P.M. Microscopy of starch: evidence of a new level of granule organization. *Carbohydr. Polym.,* **1997,** *32,* 177-191.
[http://dx.doi.org/10.1016/S0144-8617(97)00008-8]

[3]　Cave, R.A.; Seabrook, S.A.; Gidley, M.J.; Gilbert, R.G. Characterization of starch by size-exclusion chromatography: the limitations imposed by shear scission. *Biomacromolecules,* **2009,** *10*(8), 2245-2253.
[http://dx.doi.org/10.1021/bm900426n] [PMID: 19627139]

[4]　Ono, H.; Yanagisawa, M.; Isogai, A. SEC-MALS study on the molecular conformation of modified starches for papermaking. *Japan TAPPI Journal.,* **2005,** *59,* 256-264.
[http://dx.doi.org/10.2524/jtappij.59.256]

[5]   Liu, C.; Qin, Y.; Li, X.; Sun, Q.; Xiong, L.; Liu, Z. Preparation and characterization of starch nanoparticles *via* self-assembly at moderate temperature. *Int. J. Biol. Macromol.,* **2016**, *84*, 354-360.
[http://dx.doi.org/10.1016/j.ijbiomac.2015.12.040] [PMID: 26708434]

[6]   Ushiyama, T.; Shimizu, N. Microencapsulation using spray-drying: The use of fine starch solution for the wall material. *Food Sci. Technol. Res.,* **2018**, *24*, 795-801.
[http://dx.doi.org/10.3136/fstr.24.653]

[7]   Valencia, G.A.; Zare, E.N.; Makvandi, P.; Gutiérrez, T.J. Self-assembled carbohydrate polymers for food applications: Review. Compre. *Reviews Food Sci. Food Safety.,* **2019**, *18*, 2009-2024.
[http://dx.doi.org/10.1111/1541-4337.12499]

[8]   Nilsson, L. Food Hydrocolloids Separation and characterization of food macromolecules using field-flow fractionation: A review. *Food Hydrocoll.,* **2013**, *30*, 1-11.
[http://dx.doi.org/10.1016/j.foodhyd.2012.04.007]

[9]   Podzimek, S. *Light Scattering, Size Exclusion Chromatography and Asymmetric Flow Field Flow Fractionation: Powerful Tools for the Characterization of Polymers, Proteins and Nanoparticles*; John Wiley & Sons, Inc.: New Jersey, **2011**, pp. 1-36.
[http://dx.doi.org/10.1002/9780470877975]

[10]   Yoshioka, H.; Shimizu, N. Characterization of the nanoscale processing of waxy rice starch using compressed hot water. *Nihon Shokuhin Kogakkaishi,* **2014**, *15*, 95-100.
[http://dx.doi.org/10.11301/jsfe.15.95]

[11]   Buléon, A.; Colonna, P.; Planchot, V.; Ball, S. Starch granules: structure and biosynthesis. *Int. J. Biol. Macromol.,* **1998**, *23*(2), 85-112.
[http://dx.doi.org/10.1016/S0141-8130(98)00040-3] [PMID: 9730163]

[12]   Gharsallaoui, A.; Roudaut, G.; Chambin, O.; Voilley, A.; Saurel, R. Applications of spray-drying in microencapsulation of food ingredients: An overview. *Food Res. Int.,* **2007**, *40*, 1107-1121.
[http://dx.doi.org/10.1016/j.foodres.2007.07.004]

[13]   Shahidi, F.; Han, X.Q. Encapsulation of food ingredients. *Crit. Rev. Food Sci. Nutr.,* **1993**, *33*(6), 501-547.
[http://dx.doi.org/10.1080/10408399309527645] [PMID: 8216812]

[14]   Whistler, R.L.; Paschall, E.F. *Starrch: Chemistry and Technology,* 2nd ed; Academic Press: New York, **1965**, pp. 575-592.

[15]   Juna, S.; Huber, A. Characterisation of normal corn starch using asymmetrical flow field-flow fractionation. *Starke,* **2012**, *64*, 18-26.
[http://dx.doi.org/10.1002/star.201100068]

[16]   Chen, M.; Bergman, C.J. Method for determining the amylose content, molecular weights, and weight- and molar-based distributions of the degree of polymerization of amylose and fine-structure of amylopectin. *Carbohydr. Polym.,* **2007**, *69*, 562-578.
[http://dx.doi.org/10.1016/j.carbpol.2007.01.018]

[17]   Rojas, C.C.; Wahlund, K.G.; Bergenståhl, B.; Nilsson, L. Macromolecular geometries determined with field-flow fractionation and their impact on the overlap concentration. *Biomacromolecules,* **2008**, *9*(6), 1684-1690.
[http://dx.doi.org/10.1021/bm800127n] [PMID: 18537296]

[18]   van Bruijnsvoort, M.; Wahlund, K.G.; Nilsson, G.; Kok, W.T. Retention behaviour of amylopectins in asymmetrical flow field-flow fractionation studied by multi-angle light scattering detection. *J. Chromatogr. A,* **2001**, *925*(1-2), 171-182.
[http://dx.doi.org/10.1016/S0021-9673(01)01020-2] [PMID: 11519803]

[19]   Motawia, M.S.; Damager, I.; Olsen, C.E.; Møller, B.L.; Engelsen, S.B.; Hansen, S.; Øgendal, L.H.; Bauer, R. Comparative study of small linear and branched α-glucans using size exclusion chromatography and static and dynamic light scattering. *Biomacromolecules,* **2005**, *6*(1), 143-151.

[http://dx.doi.org/10.1021/bm049634e] [PMID: 15638514]

[20]   Zimm, B.H. The scattering of light and the radial distribution function of high polymer solutions. *J. Chem. Phys.,* **1948**, *16*, 1093-1099.
[http://dx.doi.org/10.1063/1.1746738]

[21]   Einstein, A. On the motion of small particles suspended in liquids at rest required by the molecular-kinetic theory of heat. *Ann. Phys.,* **1905**, *17*, 549-560.
[http://dx.doi.org/10.1002/andp.19053220806]

[22]   Herrero, E.P.; Valle, E.M.M.D.; Galán, M.A. Modelling prediction of the microcapsule size of polyelectrolyte complexes produced by atomization. *Chem. Eng. J.,* **2006**, *121*, 1-8.
[http://dx.doi.org/10.1016/j.cej.2006.04.003]

[23]   Flory, P.J.; Fox, T.G. Treatment of Intrinsic Viscosities. *J. Am. Chem. Soc.,* **1951**, *73*, 1904-1908.
[http://dx.doi.org/10.1021/ja01149a002]

[24]   Ptitsyn, O.B.; Eizner, Y.E. Hydrodynamics of polymer solutions. II. Hydrodynamic properties of macromolecules in active solvents. *Sov. Phys. Tech. Phys.,* **1960**, *4*, 1020-1036.

[25]   Hizukuri, S. Towards an understanding of the fine structure of starch molecules. *Denpun Kagaku.,* **1993**, *40*, 133-147.

[26]   Shimizu, N.; Ushiyama, T. Structure of fine waxy rice starch prepared *via* a compressed hot water process. *Food Sci. Technol. Res.,* **2018**, *24*, 795-801.
[http://dx.doi.org/10.3136/fstr.24.795]

CHAPTER 4

Major Metabolites of Certain Marketed Plant Alkaloids

Kuntal Manna*, **Bikash Debnath** and **Waikhom Somraj Singh**

Natural cum Advance Synthetic Lab, Department of Pharmacy, Tripura University (A Central University), Suryamaninagar, Tripura, India

Abstract: The archeological and historical record shows that people across Asia, Europe, and Africa used alkaloidal drugs as early as 2000 BCE. Alkaloids are heterocyclic rings consisting of at least one nitrogen atom. They are the waste products of plant metabolites and serve a wide variety of biological activities to human beings. Nicotine, cytosine, atropine, scopolamine, cocaine, catuabine, quinine, quinidine, dihydroquinine, papaverine, ephedrine, reserpine, ergotamine, caffeine, *etc.* are the most important marketed plant alkaloidal drugs and their metabolites are described in this chapter. Metabolism plays a central role in regulating the toxicity of a variety of phytochemicals. Hepatic microsomal enzymes such as monooxygenase and putative NADPH-FMN-reductase, carboxyl esterase, CYP2B6, CYP3A4, and CYP2D6 are mostly involved in the metabolism of alkaloids. This chapter will be important for future researchers.

Keywords: Cytochrome P650, Heterocyclic Ring, Hepatic Microsomal Enzymes, Marketed Alkaloids, Major Metabolites , Metabolic Pathway, Pharmacological Activities, Secondary Metabolites.

INTRODUCTION

Natural medicines provide a major source of pharmaceuticals, which we use today directly from nature or in marketed form. For the assistance of plants to survive and reproduce, they synthesize many secondary metabolites [1]. These secondary metabolites reveal biodynamic activity beneficial to both human and animal health. Alkaloids, phenols, steroids, glycosides, tannins, terpenoids, and phytoalexins are the secondary metabolites produced by the plants [2]. Among these secondary metabolites, alkaloids are considered the important ones. They are relatively modest molecules existing in plants at <10 g/kg [3]. Due to being toxic in nature, plants use alkaloids to protect themselves against harmful organi-

---

* **Corresponding author Kuntal Manna:**Department of Pharmacy, Tripura University (A Central University), Suryamaninagar, Tripura, India-799022; Tel: +91381-2379404; Fax: +91381-2374803; E-mail: k_manna2002@yahoo.com

sms. Alkaloids are constituted in the plant kingdom and mainly found in the higher plants, such as those belonging to papaveraceae, menispermaceae, ranunculaceae, leguminosae, and loganiaceae [4, 5]. Groups of nitrogen-containing compounds that may consist of one or more nitrogen atoms (within the heterocyclic ring) are called alkaloids. However, the term 'alkaloid' (alkali-like) is rather interesting as there is no definite borderline between alkaloids and naturally existing complex amines. Typical alkaloids are basic in nature and primarily acquired from plant sources [6, 7]. Alkaloidal drug metabolism can have significance due to its therapeutic effect or its toxicity. It mainly takes place in the liver and the Cytochrome P450 enzymes are involved in a vital role in metabolism [8]. This chapter mainly focuses on biological sources along with major metabolites in some important marketed alkaloid drugs.

Major Metabolites of Certain Important Marketed Alkaloids

1. Metabolite Study of Pyridine Group of Alkaloids

Nicotine 1 (Fig. (**1**) is a naturally occurring alkaloid found in many plants. Dried leaves of *Nicotiana tabacum* belonging to the family Solanaceae constitute the active source of nicotine [9]. Nicotine and its metabolites may be dangerous to the body. Hepatic enzyme Cytochrome P450 2A6 (CYP2A6), UDP-glucuronosyltransferase (UGT), and flavin-containing monooxygenase (FMO) play an active role in nicotine metabolism. Electrospray ionization and high-performance liquid chromatography/tandem-mass spectrometry (LC-MS/MS) methods are used for the determination of nicotine metabolites Fig. (**1**). Quantitatively, the most important metabolites of nicotine in mammals are the lactam derivative and cotinine. In humans, about 70– 80% of nicotine is converted into cotinine. Nicotine N'-oxide is another primary metabolite of nicotine, although only about 4–7% of nicotine consumed by active smokers is metabolized via this route. The kidney is the vital organ for the excision of nicotine [10, 11]. **Cytisine 26** Fig. (**2**) is a selective nicotinic cholinergic agonist alkaloid obtained from the seed of *Laburnum anagyroides* belonging to the family Fabaceae [12]. Astroug *et al.* (2010) described the pharmacokinetics of cytisine. They gave oral and intravenous administration to examine the cytisine pharmacokinetics profile in male and female New Zealand rabbits. After the administration of cytisine, the rabbit serum of both the sexes were collected in a specific time interval to measure the pharmacokinetic parameters using high-performance liquid chromatography (HPLC) method with ultraviolet (UV) detection. The pharmacokinetic analysis suggested a rapid but incomplete absorption of cytisine after oral administration and did not clarify any metabolites of cytisine [13]. Later Jeong *et al.* (2015) developed a liquid-chromatography mass spectrometry (LCMS) method for the pharmacokinetic study of cytisine in human plasma and urine. No metabolites were detected in plasma or urine collected in their study [14]. Further research is required for a better understanding.

Fig. (1). Metabolism pathway of nicotine (**1**), Nicotine $\Delta^{1'(5')}$ iminium ion (**2**), cotinine (**3**), Nicotine glucoronide (**4**), Nicotine N'- oxide (**5**), Nicotine isomethonium ion (**6**), 2- hydroxy nicotine (**7**), Nornicotine (**8**), Cotinine N- oxide (**9**), N'- hydroxymethyl norcotine (**10**), 5- hydroxycotinine (**11**), *Trans*-3'-hydroxycotinine (**12**), Cotinine methonium ion (13), Cotinine glucorinide (14), 4- (methylamino)-1- (3-pyridiyl)-1-butanone (**15**), Norcotine (**16**), 4-oxo-4- (3-pyridyl)-butanamide (**17**), 4-oxo-4(3-pyridyl) N-methylbutanamide (**18**), *Trans*-3'-hydroxycotinien glucoronide (**19**), 4-oxa-4(3-pyridyl)-butanoic acid (**20**), 4-hydroxy-4-(3-pyridyl butanoic acid (**21**), 5-(3-pyridyl)-tetrahydrofuran-2-one (**22**), 4-(3-pyridyl)-3-butanoic acid (**23**), 4-(3-Pyridyl)-butanoic acid (**24**), 3-Pyridylacetic acid (**25**).

2. Metabolite Study of Tropane Group of Alkaloids

Atropine 27 Fig. (**3**) is obtained from the plant *Atropa belladonna,* a perennial herb belonging to the family Solanaceae. For intoxication in nature, this drug typically provides anticholinergic effects to the body [15]. Atropine is metabolized in the liver, and 30-50% of its unchanged are excreted along with urine [16]. Chen *et al.* (2006) outlined a metabolic pathway of atropine from rat urine, based on the LC-MS/MS technique after the administration of atropine sulfate. The identified active metabolites of atropine are represented in Fig. (**3**) [17]. **Scopolamine 39** Fig. (**4**) occurs in the leaves of the *Hyoscyamus niger,* belonging to the family Solanaceae [18]. *In vivo* study of scopolamine was investigated using an extremely specific and sensitive liquid chromatography–mass spectrometry method. Various extractions (free fraction, acid and enzymatic hydrolysis) of urine (contained scopolamine) were prepared, and their comparison was carried out in the investigation of the metabolism of standard scopolamine. Scopolamine has undergone oxidative demethylation during incubation with Cytochrome P4503A (CYP3A). The major metabolites of scopolamine are norscopine, scopine, tropic acid, aponorscopolamine, aposcopolamine, and norscopolamine. The metabolic pathway of scopolamine is represented in Fig. (**4**). Less than 10% of scopolamine is excreted via urine [19, 20]. **Cocaine 58** (Fig. **5**) is a plant alkaloid isolated from the dried leaves of *Erythroxylum coca* and *Erythroxylum truxillense,* belonging to the family Erythroxylaceae [21]. The metabolism of cocaine Fig. (**5**) primarily takes place in the liver. Identified major metabolites were benzoylecgonine and ecgonine methyl ester, and minor metabolites were norcocaine, p-hydroxycocaine, m-hydroxycocaine, p-hydroxybenzoylecgonine, and m-hydroxybenzoylecgonine. Microsomal enzymes Cytochrome P450 3A4 (CYP3A4) demethylated the toxic pathway. Human carboxylesterases-1 (hCE-1), carboxylesterases-2 (hCE-2), and butyrylcholinesterase (BChE) favor the metabolism of cocaine [22, 23]. 1 to 9% of cocaine is progressively eliminated unchanged in the urine with acid [24]. **Catuabines** A (**66**), B (**67**), C (**68**), and D (**69**) Fig. (**6**) are a group of tropane alkaloids, mostly obtained from the bark of *Trichilia catigua,* belonging to the family Meliaceae [25 - 27]. The metabolism study of catuabine needs further investigation for a better understanding.

Fig. (2). Chemical structure of cytisine (**26**).

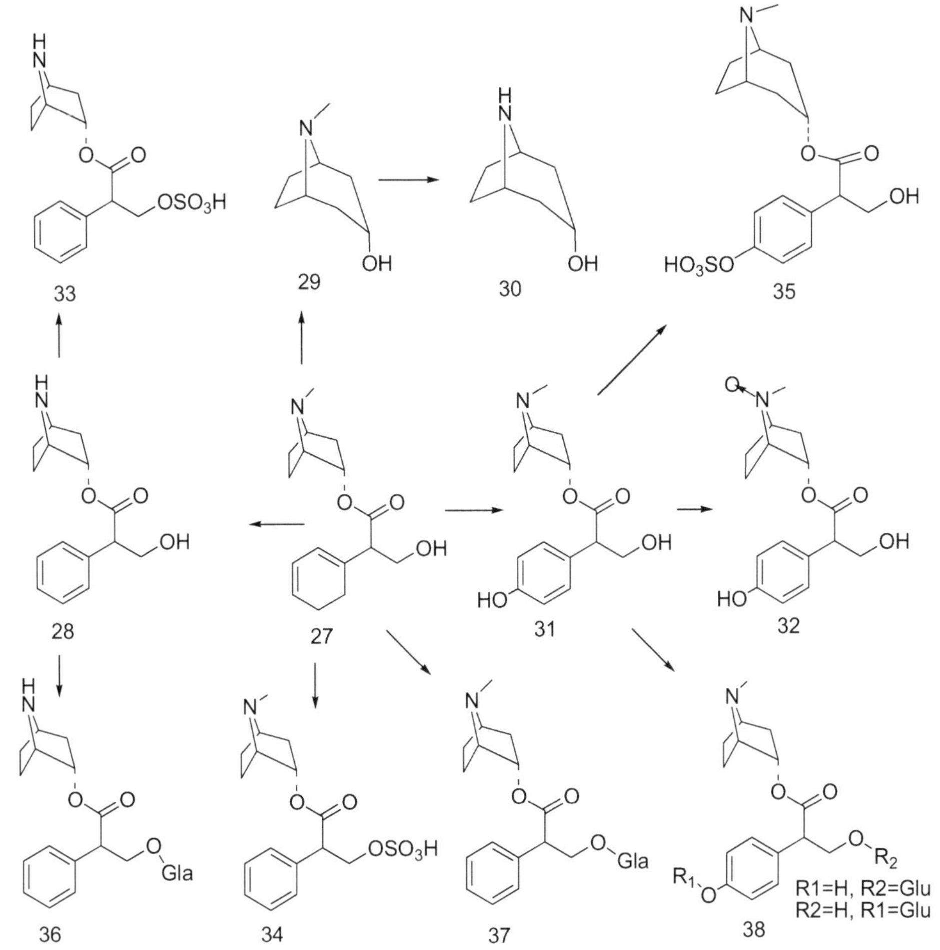

Fig. (3). Metabolism pathway of atropine (**27**), N- demethylatropine (**28**), Tropine (**29**), N- demethyltropine (**30**), *p*-hydroxyatropine (**31**), *p*-hydroxyatropine N-oxide (**32**), Sulfate conjugated N- demethylatropine (**33**), Sulfate conjugated atropine (**34**), Sulfate conjugated *p*-hydroxyatropine (**35**), Glucuronide conjugated N-demethylatropine (**36**), Glucuronide conjugated atropine (**37**), Glucuronide conjugated *p*-hydroxyatropine (**38**).

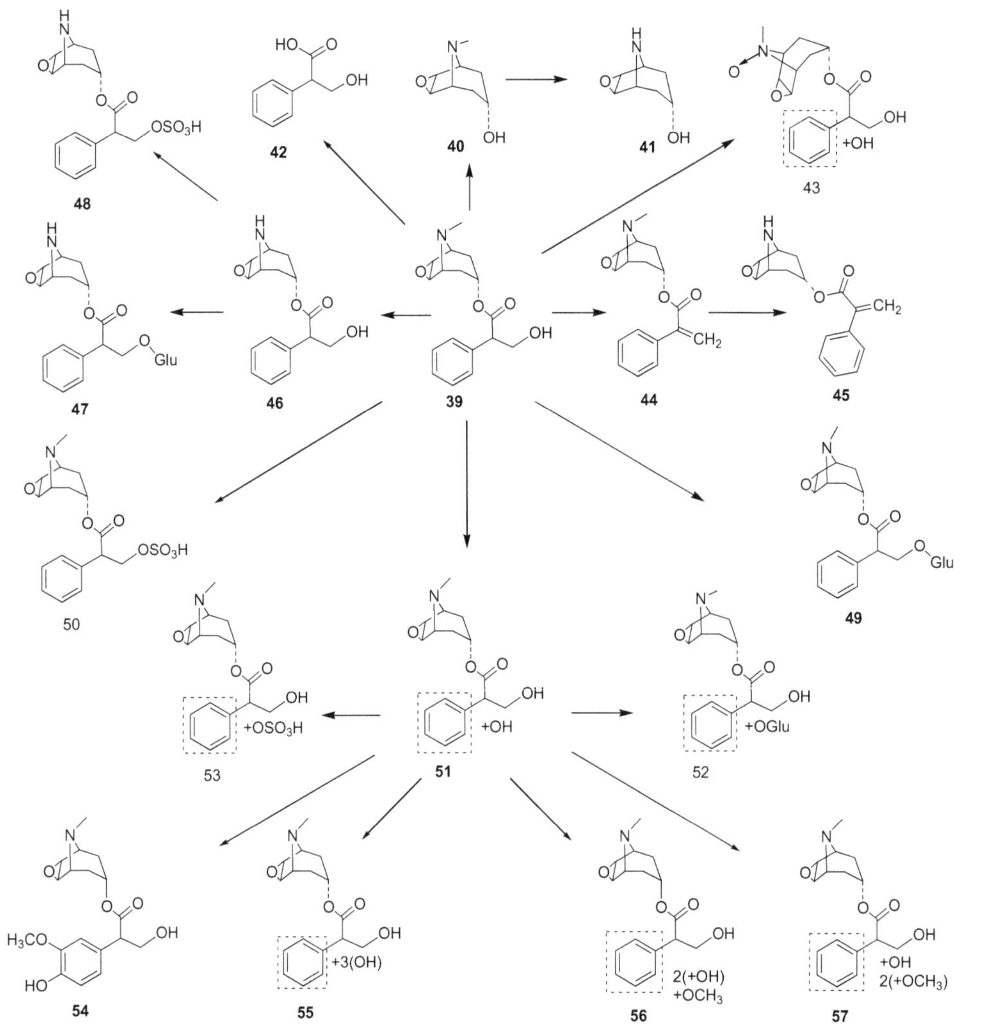

Fig. (4). Metabolism pathway of scopolamine (**39**), Scopine (**40**), Norscopine (**41**), Trapic acid (**42**), Hydrohyscopolamine N-oxide (**43**), Aposcopolamine (**44**), Aponorscopolamine (**45**), Norscopolamine (**46**), Glucoronode conjugated norscopolamine (**47**), Sulfate conjugated norscopolamine (**48**), Glucuronode conjugated scopolamine (**49**), Sulfate conjugated scopolamine (**50**), Hydroxyscopolamine (**51**), Glucuronide conjugated hydroxyscopolamine (**52**), Sulfate conjugated hydroxyscopolamine (**53**), p-hydroxyscopolamine (**54**), Trihydroxyscopolamine (**55**), Dihydroxy-methoxy scopolamine (**56**), hydroxy-dimethoxy scopolamine (**57**).

Fig. (5). Proposed metabolic pathway of Cocaine (**58**), Norcocaine (**59**), *m*-hydroxy cocaine (**60**), Ecgonine methyl ester (**61**), Benzylecgonine (**62**), *p*-hydroxycocaine (**63**), *m*-hydroxy benzylecgonine (**64**), *p*-hydroxybenzylecgonine (**65**).

Fig. (6). Chemical structure of catuabine A (**66**), catuabine B (**67**), catuabine C (**68**), catuabine D (**69**).

3. Metabolite Study of Quinoline Group of Alkaloids

Quinine 70 Fig. (7) is obtained from the bark of *Cinchona officinalis* belonging to the family Rubiaceae, which were used for the treatment of malaria. Quinine remains the first anti-malarial drug involved from as early as the 1600s [28, 29]. The active metabolite of quinine was identified in human urine as early as 1950. HPLC was used for the identification of quinine metabolites. *In vitro* studies have shown that a human liver microsomal enzymes Cytochrome P450 3A4

(CYP3A4), Cytochrome P450 1A2 (CYP1A2), and Cytochrome P450 2C19 (CYP2C19) were involved in the metabolism of quinine. The major metabolites of quinine were 2′-quininone (**71**), 3-hydoxyquinine (**72**), (10S), and (10R)-11-dihydroxydihydroquinine (**73**) Fig. (**7**). Most of the administered quinine was eliminated in the urine as a parent drug-free [30, 31]. **Quinidine** (**74**) Fig. (**8**) is obtained from the bark of *Cinchona officinalis,* belonging to the family of Rubiaceae [32]. 50 to 90% of quinidine is metabolized in the liver as hydroxylated products. Quinidine metabolites were identified by LC-MS/MS. (3S)-3-Hydroxyquinidine (**75**) and quinidine N-oxide (**76**) Fig. (**8**) remain the major products of quinidine metabolism. Quinidine 10, 11-dihydrodiol R (**77A**) and quinidine 10,11-dihydrodiol S (**77B**) Fig. (**8**) were other active metabolites of quinidine. CYP3A4 is the active microsomal enzyme involved in quinidine metabolism [33 - 35]. After 24 hours of administration of quinidine (unchanged + active metabolites), the 50% dose was excreted in the urine [36]. **Dihydroquinine 78** Fig. (**9**) is a natural impurity found in the commercial pharmaceutical formulations of quinine. The biological sources of dihydroquinine and quinine are the same. It is closely related to quinine due to cinchona alkaloids [37, 38]. The metabolism of dihydroquinine needs further study for better understanding. **Dihydroquinidine 79** (Fig. **10**) is also a cinchona alkaloid obtained from the bark of *Cinchona officinalis* belonging to the family Rubiaceae [39]. Flovat *et al.* (1989) first identified the active metabolites of dihydroquinidine using a computer program. They identified the six active metabolites of dihydroquinidine. These are (3S) - hydroxy- dihydroquinidine, 11- hydroxy- dihydroquinidine, 10-(3S) - hydroxy- dihydroquinidine, dihydroquinidine-N-oxide-1, 2'-dihydroquinidinone, and O-dihydroquinidine. The metabolic pathway of dihydroquinidine needs further research [40, 41].

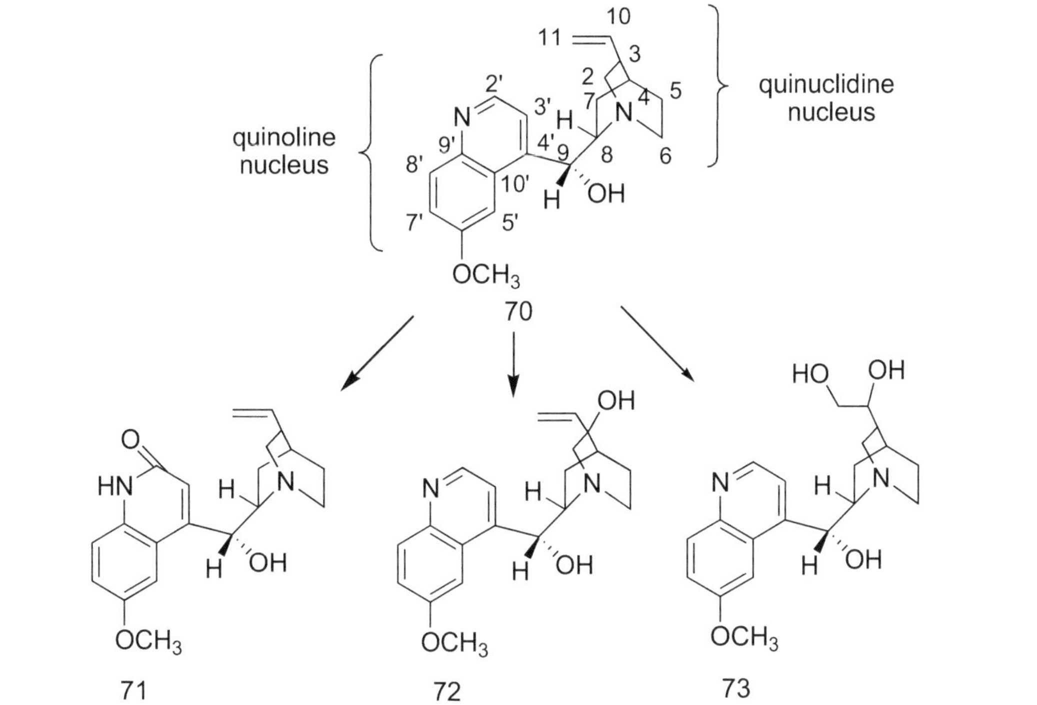

Fig. (7). Proposed metabolic pathway of quinine (**70**), 2′-quinone (**71**), 3-hydroxyquinine (**72**), 10, 11-Dihydroxydihydro quinine (**73**)

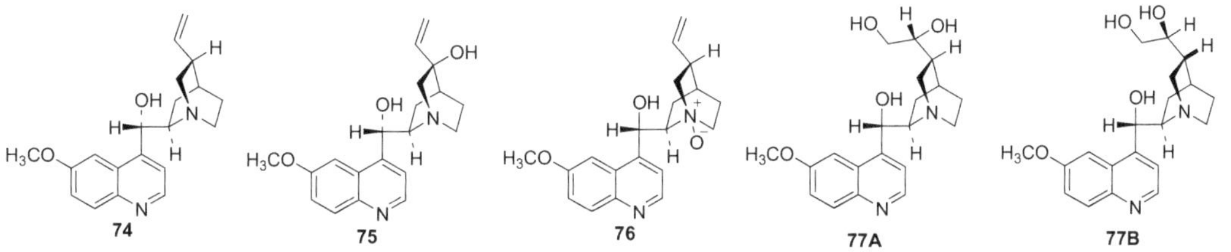

Fig. (8). Major metabolites of quinidine (**74**), (3S)-3-Hydroxyquinidine (**75**), quinidine N-oxide (**76**), Quinidine 10, 11-dihydrodiol R (**77A**) and quinidine 10, 11-dihydrodiol S (**77B**).

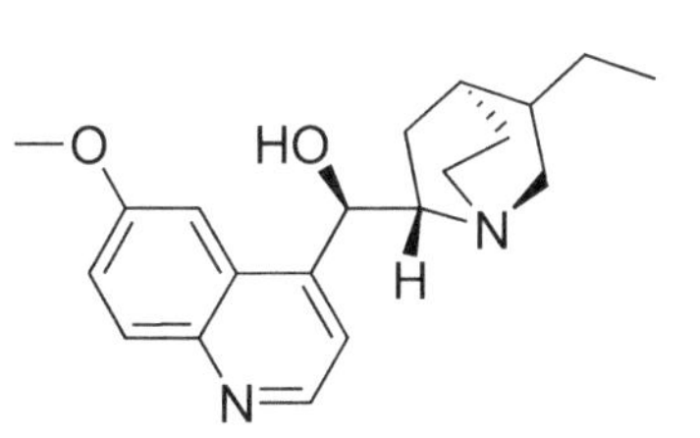

Fig. (9). Chemical structure of Dihydroquinine (**78**).

79

Fig. (10). Chemical structure of dihydroquinidine (**79**).

4. Metabolite study of Isoquinoline Group of Alkaloid

***Papaverine* 80** (Fig. **11**) is a benzylisoquinoline alkaloid, obtained from the plant *Papaver somniferum,* belonging to the family Papaveraceae [42, 43]. According to the rule of drug metabolism, the possible structures of the metabolites were speculated. A rapid liquid chromatography–electrospray ionization–ion trap mass spectrometry (LC–ESI–ITMS) was developed for the identification of papaverine and its active metabolites in rat urine. The identified metabolites are represented in Fig. (**11**) [44].

5. Metabolite study of Phenanthrene Group of Alkaloid

Codeine 95 Fig. (**12**) and **morphine 96** Fig. (**12**) are dried latex obtained by incision from the unripe capsules of *Papaver somniferum* and contain phenanthrene derivatives [45]. The principal pathways Fig. (**12**) for the efficient metabolism of codeine and morphine take place in the liver, although some of the metabolism occurred in the intestine and brain. Approximately 50-70% of the codeine is converted to codeine glucuronide by UDP-glucuronosyltransferase-2B7 (UGT2B7), 10-15% to norcodeine with the help of cytochrome P450 3A4 (CYP3A4) and 60% of morphine is glucuronidated to morphine-3-glucuronide while 5-10% of normorphine are converted from morphine. These reactions are principally catalyzed by UGT2B7 in the liver [46 - 48].

6. Metabolite Study of Phenylethylamine Group of Alkaloids

Ephedrine 105 (Fig. **13**) is an adrenergic amine present in many pharmaceutical preparations, obtained by organic synthesis or from natural sources, belonging to the genus Ephedra (Ephedraceae) like *E. vulgaris, E. siniga, E. major, E. geardiana,etc* [49]. Tseng *et al.* (2006) reported that metabolic yields studies of the ephedrine have been carried out in the human urine. They identified the metabolite products by GC-MS techniques. They also reported that ephedrine (EPH), pseudoephedrine (PEPH), phenylpropanolamine (PPA), methyl ephedrine (MEPH), and cathine are sympathomimetic amines and these obtain the primary metabolites of ephedrine. Most of the ephedrines were excreted unchanged in the urine, including EPH (40.9%), PEPH (72.2%), and PPA (59.3%). However, unchanged MEPH was estimated at 15.5% in the urine. Enzyme kinetic for ephedrine metabolism needs further study. The metabolism pathway of ephedrine is represented in Fig. (**13**) [50].

Fig. (11). The proposed *in vivo* metabolic pathways of papaverine (**80**) in rat, Dimethyl metabolites of papavarine (**81**), De-dimethyl metabolites of papaverine (**82**), De-trimethyl product of papaverine (**83**), Mono-hydroxylation product of first metabolites (**84**), Mono-hydroxylation product of second metabolites (**85**), Mono-hydroxylation product of third metabolites (**86**), and Mono-hydroxylation product of papavarine (**87**), Sulphated conjugate of first metabolites (**88**), Glucuronide conjugate of first metabolites (**89**), Sulphated conjugate of second metabolites (**90**), Glucuronide conjugate of second metabolites (**91**), Glucuronide conjugate of fourth metabolites (**92**), Glucuronide conjugate of fifth metabolites (**93**), Glucuronide conjugate of seventh metabolites (**94**).

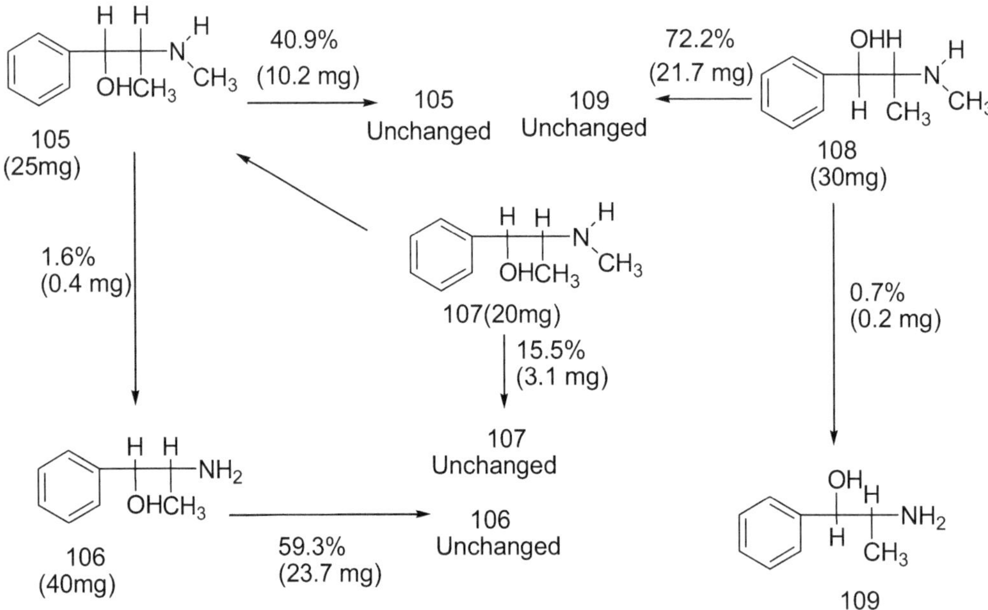

Fig. (12). Metabolic pathways for codeine (**95**) and morphine (**96**), Hydrocodone (**97**), Hydromorphine (**98**), Norcodine (**99**), Norcodine glucoronide (**100**), Clodine glocoronide (**101**), Morphine 3-glucoronide (**102**), Normorphine (**103**), Normorphine glucoronide (**104**).

Fig. (13). Metabolic pathways for ephedrine (EPH) (**105**), phenylpropanolamine (PPA) (**106**), methylephedrine (MEPH) (**107**), pseudoephedrine (PEPH) (**108**), and cathine (**109**).

7. Metabolite Study of Indole Group of Alkaloids

Reserpine 110 (Fig. **14**) is a naturally occurring drug obtained from a genus of Rauwolfia. Most of the time, reserpine is isolated from the root of *Rauwolfia serpentine* and *Rauwolfia vomitoria* [51]. Hepatic metabolism accounts for less than 50% of the elimination of reserpine, with the remainder being eliminated in the feces and some unmetabolized and active metabolites in the urine. In man, the major metabolites of reserpine are methyl reserpate and 3, 4, 5-trimethoxybenzoic acid. The other important metabolite is reserpic acid. The metabolic pathway and the responsible enzyme for the metabolism of reserpine need further study. Major metabolites of reserpine are represented in Fig. (**14**) [52, 53]. **Ergotamine 114** (Fig. **15**) is a fungal infection that has naturally or intentionally infected rye (*Claviceps purpurea*) and other plants since the beginning of farming [54]. The liver metabolizes an undefined pathway's ergotamine, and metabolites are excreted in the bile [55]. The major metabolites of ergotamine are monohydroxylated and dihydroxylated [56]. Dellafiora *et al.* (2015) described ten monohydroxylated and dihydroxylated classes of ergotamine metabolites Fig. (**15**) [57]. Cytochrome P4503A (CYP3A) supports the metabolism of ergotamine. The putative name of these ten metabolites needs further study [58]. **Yohimbine 125** (Fig. **16**) is an indole alkyl-amine alkaloid, which is chemically identical to reserpine. It is isolated from the bark of *Pausinystalia johimbe* belonging to the family Rubiaceae [59]. Extensive metabolism of yohimbine appeared to undergo an organ (liver or kidney) with significant flow. A normal-phase high-performance liquid chromatography (HPLC) and liquid chromatography-mass spectrometry (LC/MS) methods were used simultaneously for the determination of yohimbine and its primary metabolites Fig. (**16**). 10-hydroxy-yohimbine **126** (Fig. **16**) and 11-hydroxy-yohimbine **127** (Fig. **16**) are the two major metabolites of yohimbine, which was investigated on a healthy human subject. The metabolic pathway of yohimbine needs further study for a better understanding [60 - 62]. Cytochrome P450 2D6 (CYP2D6) and cytochrome P450 3A4 (CYP3A4) are the two important genotypes of P450 involved in the metabolism of yohimbine [63]. Naturally, **vinblastine 128** Fig. (**17**) is extracted from the pink periwinkle plant, *Catharanthus roseus,* belonging to the family Apocynaceae [64]. Tellingen *et al.* (1993) used ion-exchange normal-phase liquid chromatography with fluorescence detection and liquid-liquid extraction methods for the analysis of vinblastine and its metabolites. They revealed that only 18% of the dose was excreted as an unchanged drug and 19% into three active metabolites. In these three metabolites, O^4-deacetylvinblastine is the first metabolic product that was identified, and the rest of the two metabolites are still unknown. The metabolic pathway of vinblastine needs further research for a better understanding [65]. Cytochrome P450 3A4 (CYP3A4) is the active human liver cytochrome enzyme that plays a key role in vinblastine metabolism [66]. Naturally, **vincristine 129** Fig. (**18**) is extracted from the *Catharanthus roseus*, pink periwinkle plant, belonging to the family Apocynaceae [64]. An analytical method electrospray ionization and high-performance liquid chromatography/tandem mass spectrometry (LC/ESI-MS/MS) was developed for the identification of active metabolites of vincristine. *In vitro* and *in vivo* data showed that cytochrome P450 3A4 (CYP3A4) and P450 3A5 (CYP3A5) were the human liver cytochrome enzymes involved in the metabolism of vincristine resulting in a substantial loss of the parent drug vincristine and formation of the one unknown major metabolite M1 (Metabolite 1) and two unknown minor metabolites, M2 (Metabolite 2) and M3 (Metabolite 3) [67 - 69]. **Serotonin** (5-hydroxytryptamine or 5-HT) **130** Fig. (**19**) is acquired from the legume of *Mucuna pruriens,* belonging to the family Fabaceae [70]. McIsaac and Page (1959) identified the serotonin metabolites by the administration of C^{14}--hydroxytryptamine in the exposed rat and rabbit. They collected the exposed rat and rabbit feces in a specific time interval and identified five metabolites. These are 5-hydroxyindoleacetic acid, 5-hydroxytryptamine glucuronide, 5-hydroxytryptamine, an Unknown oxidation product, N-5-hydroxytryptamine [71].

Fig. (14). Major metabolites of reserpine (**110**), Methyl reserpine (**111**), 3, 4, 5-trimethoxybenzoic acid (**112**), Reserpic acid (**113**).

Fig. (15). Major metabolites of Ergotamine (**114**), Metabolites1 (**115**), Metabolites 2 (**116**), Metabolites 3 (**117**), Metabolites 4 (**118**), Metabolites 5 (**119**), Metabolites 6 (**120**), Metabolites 7 (**121**), Metabolites 8 (**122**), Metabolites 9 (**123**), and Metabolites 10 (**124**).

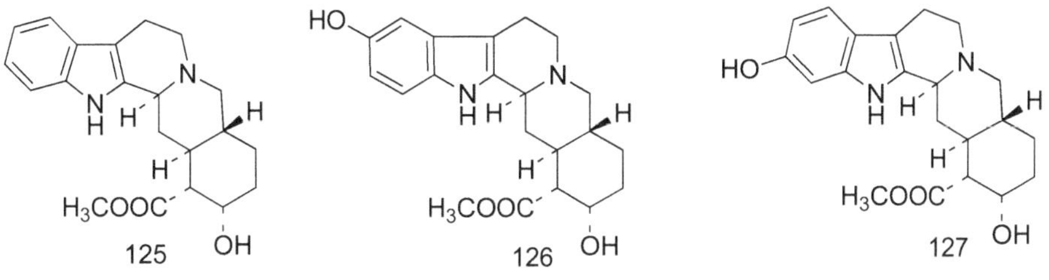

Fig. (16). Major Metabolites of Yohimbine (**125**), 10-hydroxyyohimbine (**126**), 11-hydroxyyohimbine (**127**).

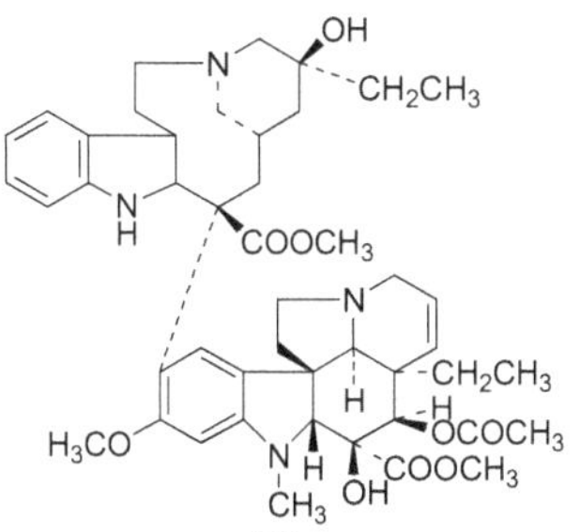

128

Fig. (17). Chemical structure of vinblastine (**128**).

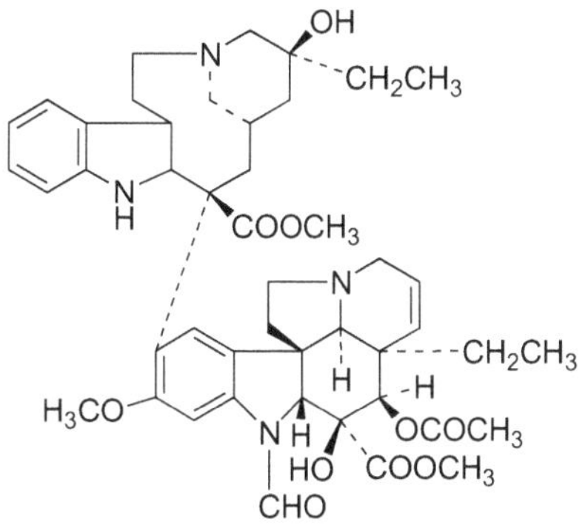

129

Fig. (18). Chemical structure of vincristine (**129**).

130

Fig. (19). Chemical structure of serotonin (**130**).

8. Metabolite Study of Purine Group of Alkaloid

Caffeine 131 Fig. (**20**) is a purine alkaloid and odorless white powder that possesses unpleasant taste. It is naturally found in the *Theobroma cocoa* (seeds) (family: Malvaceae) and *Thea sinensis* (leaves) (family: Theaceae) [72]. With 3% or less excreted unchanged in the urine, caffeine almost completely metabolized. 70-80% of the active metabolites (paraxanthine, theophylline, and theobromine) are produced via the demethylation process. 1,3,7-trimethyluric acid is obtained as an important metabolite of caffeine and is produced through the direct oxidation process. The primary identified metabolites of theophylline are methyluric acid. On the other hand, theobromine undergoes demethylation and oxidation reactions. In the demethylation process, metabolites are identified as 7-methylxanthine, 7-methyluric acid, 3-methyl xanthine and 3- methyluric acid. The identified metabolites under the oxidation process were 3, 7-dimethyl uric acid and 6-amino-5-(N-formylamino)-1-methyl uracil. For the primary metabolism of caffeine, the enzyme Cytochrome P450 1A2 (CYP1A2) is responsible for more than 95%. Metabolism pathway of caffeine is represented in Fig. (**20**) [73 - 75].

9. Metabolite Study of Imidazole Group of Alkaloid

Pilocarpine 143 Fig. (**21**) is isolated from the leaves of *Pilocarpus microphyllus* belonging to the family Rutaceae [76]. In humans, hydrolysis and hydroxylation are the major pathways Fig. (**21**) involved in pilocarpine metabolism. The 3-hydroxypilocarpine was identified as an active metabolite by hydroxylation and cytochrome P450 2A6 (CYP2A6) conciliated the metabolism. Pilocarpic acid is another active metabolite produced by the hydrolysis of pilocarpine [77, 78].

Fig. (20). Metabolic pathway for caffeine (**131**); Paraxanthine/ 1,7- Dimethylxanthine (**132**), Theophylline/ 1,3- Dimethylxanthine (**133**), Theobromine/ 3,7- Dimethylxanthine (**134**), 1,3,7- Trimethyluric acid (**135**), Methyl uric acid/ 1,3- Dimethyl uric acid (**136**), Methyl xanthine (**7**), 2-Methylxanthine (**138**), 3,7- Dimethyl uric acid (**139**), 7- Methyl uric acid (**140**), 3- Methyl uric acid (**141**), 6- Amino 5-(N- Methylformylamino)-1-methyl uracil (**142**).

Fig. (21). Metabolic pathway of Pilocarpine (**143**); Pilocarpic acid (**144**), 3-hydroxypilocarpic acid (**145**).

10. Metabolite Study of Terpenoid Group of Alkaloids

Capsaicin 146 Fig. (**22**) is primarily found in the fruit of the Capsicum genus like *Capsicum annum* and *Capsicum frutescens,* belonging to the family Solanaceae [79]. Different types of cytochrome P450 (CYP1A1, 1A2, 2B6, 2C8, 2C9, 2C19, 2D6, 2E1, and 3A4) enzymes are involved in the metabolism of capsaicin. A unique combination of liquid chromatography-mass spectrometry (LC/MS), liquid chromatography-tandem mass spectrometry (LC/MS/MS), and liquid chromatography and nuclear magnetic resonance (LC/NMR) was predominantly used for the identification of several metabolites Fig. (**22**) of capsaicin. The metabolism of capsaicin was found similar in the microsomes of a rat, dog, and human [80, 81]. **Choline 153** Fig. (**23**) is found in a small number of plant foods (vegetables, nuts, seeds, and whole grains). According to the Food and Nutrition Board of India, choline is essential for the human diet [82]. Choline is the precursor of various metabolites. The intracellular routing of the choline to its various metabolic pathways is phosphorylation, oxidation, and acetylation to cell and tissue. Choline metabolism and choline-derived metabolites can undergo extensive alterations because of a malignant transformation. The major metabolites of choline are acetylcholine, betanine, glycerophosphorylcholine, sphingomyelin, and phosphatidylcholine. Major metabolites of choline are represented in Fig. (**23**) [82, 83]. Summarization of the generic name, brand name, major metabolites, and biological activity of marketed alkaloid drugs is represented in Table **1**.

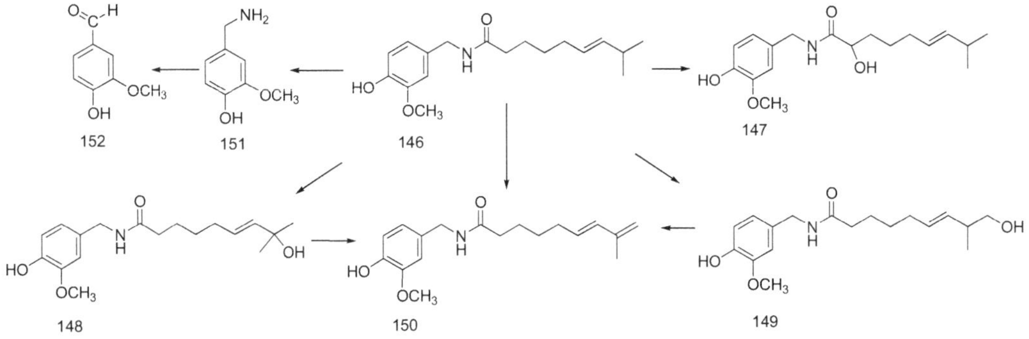

Fig. (22). Metabolic pathway of Capsaicin (**146**); Hydroxy capsaicin (**147**), 16-Hydroxy capsaicin (**148**) 17-Hydroxy capsaicin (**149**), 16, 17-Dehydro capsaicin (**150**), vanillylamine (**151**), vanillin (**152**).

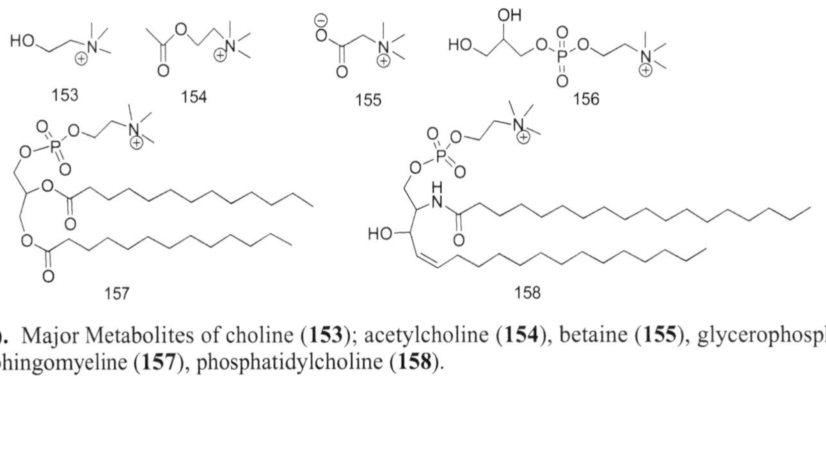

Fig. (23). Major Metabolites of choline (**153**); acetylcholine (**154**), betaine (**155**), glycerophosphorylcholine (**156**), sphingomyeline (**157**), phosphatidylcholine (**158**).

Table 1. Summarization of marketed plant alkaloids mentioned in the chapter.

Name of Marketed Alkaloids (Generic Name)	Brand Name	Important Pharmacological Activity	Major Metabolites
Nicotine	Habitrol, Nicoderm C-Q	Behavior-modifying effects	Cotinine, Nicotine N'-oxide and cotinine N'-oxide
Cytisine	Desmoxan, Tabex	Improved risk groups of smokers with health problems on the part of the cardiovascular and respiratory systems	ND
Atropine	Sal-Tropine, AtroPen, Atreza	Anti-cholinergic effects	Noratropine, atropine-N-oxide (equatorial isomer), tropine and tropic acid
Scopolamine	Transderm-Scop, scopace, maldemar	Central sedative, antiemetic, and amnestic effects	Norscopine, scopine, tropic acid, aponorscopolamine
Cocaine	Goprelto	CNS stimulant, restorative and also in convulsion	Benzoylecgonine (BE) and ecgonine methyl ester (EME)
Catuabine	Herbal era Catuaba 30, Hawaiian Herbals Catuaba Bark Capsule	Improved nutritional balance and antidepressant-like effects	ND
Quinine	Qualaquin, QM-260, Quinamm	Antimalarial, antipyretic, anti-smallpox, analgesic and anti-inflammatory	3-hydoxyquinine, 2'-quininone, (10S) and (10R)-11-dihydroxydihydroquinine
Quinidine	Quin-G, Cardioquin, Quinora, Quinidex Extentabs, Quinaglute Dura-Tabs, Quin-Release	Antimalarial	3-Hydroxyquinidine and quinidine N-oxide
Dihydroquinine	Hidrospot, Pigmentasa, Inhibin	Inhibit the actions of the parasympathetic nervous system	ND
Dihydroquinidine	Austacute, Idrochinidina Lirca, Lentoquine, Sérécor LP	Dihydroquinine and dihydroquinidine have the same degree of activity	(3S)- hydroxy- dihydroquinidine (3-OH-DHQ), 11- hydroxy- dihydroquinidine (11-OH-DHQ), 10-(3S)- hydroxy- dihydroquinidine (10-OH-DHQ)
Papaverine	Papacon, Pavabid Plateau, Pavagen, Pavacot	Relaxation of the tonus of all smooth muscle	De-methoxyl, hydroxyl, glucuronide, and sulfate conjugated metabolites are identified

(Table 1) cont.....

Name of Marketed Alkaloids (Generic Name)	Brand Name	Important Pharmacological Activity	Major Metabolites
Morphine	Arymo ER, Kadian, MorphaBond ER, MS Contin	Opioid analgesic	Morphine-3-glucuronide, normorphine
Codeine	Fioricet with Codeine, Phrenilin with Caffeine and Codeine	Opioid analgesic	Codeine glucuronide, norcodeine
Ephedrine	Akovaz, Corphedra	Use as a sympathetic stimulation directly act on α and β receptor	Ephedrine, pseudoephedrine, phenylpropanolamine, methylephedrine, cathine
Reserpine	Diupres-250, Diupres-500	Antihypertensive activity	Methylreserpate and trimethoxybenzoic acid
Ergotamine	Ergomar	Increase the motor activity of the uterus	Ten monohydroxylated and dihydroxylated classes metabolites are identified but their putative name has not described
Yohimbine	Yohimbe, Yocon, Erex, Yohimex	Increase parasympathetic (cholinergic) and decrease sympathetic (adrenergic) activity, developed male sexual performance	10-hydroxy-yohimbine, 11-hydroxy-yohimbine
Vinblastine	Velban	Anti-cancer, Hodgkin's sickness activity	O^4-deacetylvinblastine and two unknown metabolites
Vincristine	Oncovin, Vincasar PFS	Antineoplastic activity	Three unknown metabolites
Serotonine	Sergotonine	Mucosal signaling molecule	5-hydroxyindoleacetic acid, 5-hydroxyindoleaceturic acid, 5-hydroxytryptamine glucuronide, 5-hydroxytryptamine
Caffeine	Vivarin, Alert, Lucidex, Cafcit	Analgesics, anorectants and CNS stimulant	Paraxanthine, Theophylline, and theobromine, 1,3,7 trimethyluric acid
Pilocarpine	Salagen	Direct cholinergic properties of pilocarpine stimulate the parasympathetic system (bladder, tear ducts, sudoriferous and salivary glands)	Pilocarpic acid, 3-hydroxypilocarpine

(Table 1) cont.....

Name of Marketed Alkaloids (Generic Name)	Brand Name	Important Pharmacological Activity	Major Metabolites
Capsaicin	Zostrix, Icy Hot PM Patch, Qutenza, Capsin	Expressed in the sensory neurons	Vanillylamine, vanillin, 16-hydroxycapsaicin, 17-hydroxycapsaicin, and 16,17-dehydrocapsaicin
Choline	Choline Magnesium Trisalicylate, Tricosal, Trilisate, CMT	Choline is a constituent of cell and mitochondrial membranes and of the neurotransmitter acetylcholine	Acetylcholine, betanine, glycerophosphorylcholine, sphingomyelin, phosphatidylcholine

**https://www.drugs.com/ and https://www.drugbank.ca/ provides the data for identified the marketed alkaloidal drugs.
*ND- not detected.

CONCLUSION

A group of alkaloids such as pyridine, tropane, quinoline, isoquinoline, phenanthrene, phenylethylamine, purine, imidazole, and terpenoid is a unique source of important pharmacologically active led molecules and useful medicines available in the market. Most of the alkaloid drugs were metabolized with the help of hepatic microsomal enzymes like monooxygenase and putative NADPH-FMN-reductase, carboxylesterase, CYP2B6, CYP3A4, and CYP2D6. Scientists developed various techniques like HPLC, LC/ESI-MS/MS, LC–ESI–ITMS, *etc.*, which were used for the determination of active metabolites. Even now, researchers are facing a challenge to identify the active metabolites and enzymes responsible for the metabolism of alkaloids. However, some alkaloids along with their metabolic pathways are published, but still, some alkaloids need further study for a better understanding.

CONSENT FOR PUBLICATION

Not applicable.

CONFLICT OF INTEREST

The authors confirm that this chapter content has no conflict of interest.

ACKNOWLEDGEMENTS

The authors are grateful for the e-resources provided by Tripura University (A Central University), Suryamaninagar, Tripura, India-799022. We are also grateful to the University Grant Commission (UGC), Govt. of India, for providing Startup Grant [F. 20-2 (24)/ 2012 (BSR)] for newly recruited faculty and as financial support. We also acknowledge the Department of Science and Technology (DST),

Govt. of India for awarding DST Fast Track Scheme [NO. SB/FT/CS-150/ 2012].

REFERENCES

[1] Jing, H.; Liu, J.; Liu, H.; Xin, H. Histochemical investigation and kinds of alkaloids in leaves of different developmental stages in *Thymus quinquecostatus*. *ScientificWorldJournal*, **2014**, *2014*839548
[http://dx.doi.org/10.1155/2014/839548] [PMID: 25101324]

[2] Arnason, J.T.; Catling, P.M.; Small, E.; Dang, P.T.; Lambert, J.D.H. Biodiversity & Health: Focusing Research to Policy *Proceeding of the international symposium,* **2005**,

[3] Acamovic, T.; Stewart, C.S.; Pennycott, T.W. *Poisonous Plants and Related Toxins*; CABI Wallingford: UK, **2004**.
[http://dx.doi.org/10.1079/9780851996141.0000]

[4] Aniszewski, T. *Alkaloids-Secrets of Life: Alkaloids Chemistry, Biological Significance, Applications and Ecological Role*; Elsevier: Amsterdam, The Netherlands, **2007**.

[5] Debnath, B.; Singh, W.S.; Das, M.; Goswam, S.; Singh, M.K.; Maiti, D.; Manna, K. Role of plant alkaloids on human health: A review of biological activities. *Mater. Today Chem.,* **2018**, *9*, 56-72.
[http://dx.doi.org/10.1016/j.mtchem.2018.05.001]

[6] Debnath, B.; Uddin, M.J.; Patari, P.; Das, M.; Maiti, D.; Manna, K. Estimation of alkaloids and phenolics of five edible cucurbitaceous plants and their antibacterial activity. *Int. J. Pharm. Pharm. Sci.,* **2015**, *7*, 223-227.

[7] Pan, S.Y.; Zhou, S.F.; Gao, S.H.; Yu, Z.L.; Zhang, S.F.; Tang, M.K.; Sun, J.N.; Ma, D.L.; Han, Y.F.; Fong, W.F. Ko, K.M. New Perspectives on How to Discover Drugs from Herbal Medicines: CAM's Outstanding Contribution to Modern Therapeutics. *Evid. Based Complement. Alternat. Med.,* **2013**, 1-25.

[8] Gasser, R. *Importance of drug metabolism in drug discovery and development. In Molecular and Applied Aspects of Oxidative Drug Metabolizing Enzymes*; Springer: Boston, **1999**, pp. 183-193.

[9] Benowitz, N.L. Pharmacology of nicotine: addiction, smoking-induced disease, and therapeutics. *Annu. Rev. Pharmacol. Toxicol.,* **2009**, *49*, 57-71.
[http://dx.doi.org/10.1146/annurev.pharmtox.48.113006.094742] [PMID: 18834313]

[10] Jacob, P., III; Yu, L.; Duan, M.; Ramos, L.; Yturralde, O.; Benowitz, N.L. Determination of the nicotine metabolites cotinine and trans-3'-hydroxycotinine in biologic fluids of smokers and non-smokers using liquid chromatography-tandem mass spectrometry: biomarkers for tobacco smoke exposure and for phenotyping cytochrome P450 2A6 activity. *J. Chromatogr. B Analyt. Technol. Biomed. Life Sci.,* **2011**, *879*(3-4), 267-276.
[http://dx.doi.org/10.1016/j.jchromb.2010.12.012] [PMID: 21208832]

[11] Gorrod, J.W.; Wahren, J. *Nicotine and related alkaloids absorption, distribution, metabolism, and excretion*; Springer- Science Business Media, **1993**, pp. 185-187.
[http://dx.doi.org/10.1007/978-94-011-2110-1]

[12] https://www.drugs.com/dict/cytisine.html

[13] Astroug, H.; Simeonova, R.; Kassabova, L.V.; Danchev, N.; Svinarov, D. Pharmacokinetics of cytisine after single intravenous and oral administration in rabbits. *Interdiscip. Toxicol.,* **2010**, *3*(1), 15-20.
[http://dx.doi.org/10.2478/v10102-010-0003-5] [PMID: 21217866]

[14] Jeong, S.H.; Newcombe, D.; Sheridan, J.; Tingle, M. Pharmacokinetics of cytisine, an α4 β2 nicotinic receptor partial agonist, in healthy smokers following a single dose. *Drug Test. Anal.,* **2015**, *7*(6), 475-482.
[http://dx.doi.org/10.1002/dta.1707] [PMID: 25231024]

[15] Çaksen, H.; Odabaş, D.; Akbayram, S.; Cesur, Y.; Arslan, S.; Üner, A.; Öner, A.F. Deadly nightshade

(Atropa belladonna) intoxication: an analysis of 49 children. *Hum. Exp. Toxicol.,* **2003**, *22*(12), 665-668.
[http://dx.doi.org/10.1191/0960327103ht404oa] [PMID: 14992329]

[16]   http://ratguide.com/meds/cardiovascular_drugs/atropine_sulfate.php

[17]   Chen, H.; Chen, Y.; Du, P.; Han, F.; Wang, H.; Zhang, H. Sensitive and specific liquid chromatographic-tandem mass spectrometric assay for atropine and its eleven metabolites in rat urine. *J. Pharm. Biomed. Anal.,* **2006**, *40*(1), 142-150.
[http://dx.doi.org/10.1016/j.jpba.2005.06.027] [PMID: 16087309]

[18]   Lewis, W.H.; Elvin-Lewis, M.P. *Medical botany: plants affecting human health*; John Wiley & Sons, **2003**.

[19]   Chen, H.; Chen, Y.; Wang, H.; Du, P.; Han, F.; Zhang, H. Analysis of scopolamine and its eighteen metabolites in rat urine by liquid chromatography-tandem mass spectrometry. *Talanta,* **2005**, *67*(5), 984-991.
[http://dx.doi.org/10.1016/j.talanta.2005.04.026] [PMID: 18970269]

[20]   Chen, H.; Chen, Y.; Du, P.; Han, F. Liquid chromatography-electrospray ionization ion trap mass spectrometry for analysis of in vivo and in vitro metabolites of scopolamine in rats. *J. Chromatogr. Sci.,* **2008**, *46*(1), 74-80.
[http://dx.doi.org/10.1093/chromsci/46.1.74] [PMID: 18218192]

[21]   Plowman, T. The ethnobotany of coca (Erythroxylum spp., Erythroxylaceae). **1984**.

[22]   Kolbrich, E.A.; Barnes, A.J.; Gorelick, D.A.; Boyd, S.J.; Cone, E.J.; Huestis, M.A. Major and minor metabolites of cocaine in human plasma following controlled subcutaneous cocaine administration. *J. Anal. Toxicol.,* **2006**, *30*(8), 501-510.
[http://dx.doi.org/10.1093/jat/30.8.501] [PMID: 17132243]

[23]   Sun, H.; Pang, Y.P.; Lockridge, O.; Brimijoin, S. Re-engineering butyrylcholinesterase as a cocaine hydrolase. *Mol. Pharmacol.,* **2002**, *62*(2), 220-224.
[http://dx.doi.org/10.1124/mol.62.2.220] [PMID: 12130672]

[24]   http://www.inchem.org/documents/pims/pharm/pim139e.htm

[25]   Campos, M.M.; Fernandes, E.S.; Ferreira, J.; Santos, A.R.; Calixto, J.B. Antidepressant-like effects of Trichilia catigua (Catuaba) extract: evidence for dopaminergic-mediated mechanisms. *Psychopharmacology (Berl.),* **2005**, *182*(1), 45-53.
[http://dx.doi.org/10.1007/s00213-005-0052-1] [PMID: 15991001]

[26]   Zanolari, B.; Guilet, D.; Marston, A.; Queiroz, E.F.; Paulo, Mde.Q.; Hostettmann, K. Tropane alkaloids from the bark of Erythroxylum vacciniifolium. *J. Nat. Prod.,* **2003**, *66*(4), 497-502.
[http://dx.doi.org/10.1021/np020512m] [PMID: 12713400]

[27]   https://en.wikipedia.org/wiki/Catuabine

[28]   Achan, J.; Talisuna, A.O.; Erhart, A.; Yeka, A.; Tibenderana, J.K.; Baliraine, F.N.; Rosenthal, P.J.; D'Alessandro, U. Quinine, an old anti-malarial drug in a modern world: role in the treatment of malaria. *Malar. J.,* **2011**, *10*, 144.
[http://dx.doi.org/10.1186/1475-2875-10-144] [PMID: 21609473]

[29]   http://www.newworldencyclopedia.org/entry/Quinine

[30]   Mirghani, R.A.; Hellgren, U.; Bertilsson, L.; Gustafsson, L.L.; Ericsson, O. Metabolism and elimination of quinine in healthy volunteers. *Eur. J. Clin. Pharmacol.,* **2003**, *59*(5-6), 423-427.
[http://dx.doi.org/10.1007/s00228-003-0637-8] [PMID: 12920491]

[31]   Babalola, C.P.; Kotila, O.A.; Dixon, P.A.; Oyewo, A.E. Disposition of quinine and its major metabolite, 3-hydroxyquinine in patients with liver diseases. *Res. Pharm. Biotech.,* **2011**, *3*, 25-29.

[32]   White, N.J.; Looareesuwan, S.; Warrell, D.A.; Chongsuphajaisiddhi, T.; Bunnag, D.; Harinasuta, T. Quinidine in falciparum malaria. *Lancet,* **1981**, *2*(8255), 1069-1071.

[http://dx.doi.org/10.1016/S0140-6736(81)91275-7] [PMID: 6118523]

[33]   Rakhit, A.; Holford, N.H.; Guentert, T.W.; Maloney, K.; Riegelman, S. Pharmacokinetics of quinidine and three of its metabolites in man. *J. Pharmacokinet. Biopharm.,* **1984**, *12*(1), 1-21.
[http://dx.doi.org/10.1007/BF01063608] [PMID: 6747817]

[34]   Nielsen, T.L.; Rasmussen, B.B.; Flinois, J.P.; Beaune, P.; Brøsen, K. In vitro metabolism of quinidine: the (3S)-3-hydroxylation of quinidine is a specific marker reaction for cytochrome P-4503A4 activity in human liver microsomes. *J. Pharmacol. Exp. Ther.,* **1999**, *289*(1), 31-37.
[PMID: 10086984]

[35]   Ngui, J.S.; Tang, W.; Stearns, R.A.; Shou, M.; Miller, R.R.; Zhang, Y.; Lin, J.H.; Baillie, T.A. Cytochrome P450 3A4-mediated interaction of diclofenac and quinidine. *Drug Metab. Dispos.,* **2000**, *28*(9), 1043-1050.
[PMID: 10950847]

[36]   Lowenthal, D.T. Pharmacokinetics of propranolol, quinidine, procainamide and lidocaine in chronic renal disease. *Am. J. Med.,* **1977**, *62*(4), 532-538.
[http://dx.doi.org/10.1016/0002-9343(77)90411-9] [PMID: 851126]

[37]   Brembilla-Perrot, B.; Jacquemin, L.; Beurrier, D. Relationships between heart rate variability and antiarrhythmic effects of hydroquinidine. *Cardiovasc. Drugs Ther.,* **1997**, *11*(3), 493-498.
[http://dx.doi.org/10.1023/A:1007709808576] [PMID: 9310279]

[38]   Nontprasert, A.; Pukrittayakamee, S.; Kyle, D.E.; Vanijanonta, S.; White, N.J. Antimalarial activity and interactions between quinine, dihydroquinine and 3-hydroxyquinine against Plasmodium falciparum in vitro. *Trans. R. Soc. Trop. Med. Hyg.,* **1996**, *90*(5), 553-555.
[http://dx.doi.org/10.1016/S0035-9203(96)90320-X] [PMID: 8944272]

[39]   Curd, F.H.; Davey, D.G.; Rose, F.L. Studies on Synthetic Antimalarial Drugs: II.—. *General Chemical Considerations. Ann. Trop. Med. Parasitol.,* **1945**, *39*, 157-164.
[http://dx.doi.org/10.1080/00034983.1945.11685229] [PMID: 21013244]

[40]   Ebert, S.N.; Liu, X.K.; Woosley, R.L. Female gender as a risk factor for drug-induced cardiac arrhythmias: evaluation of clinical and experimental evidence. *J. Womens Health,* **1998**, *7*(5), 547-557.
[http://dx.doi.org/10.1089/jwh.1998.7.547] [PMID: 9650155]

[41]   Ching, M.S.; Blake, C.L.; Ghabrial, H.; Ellis, S.W.; Lennard, M.S.; Tucker, G.T.; Smallwood, R.A. Potent inhibition of yeast-expressed CYP2D6 by dihydroquinidine, quinidine, and its metabolites. *Biochem. Pharmacol.,* **1995**, *50*(6), 833-837.
[http://dx.doi.org/10.1016/0006-2952(95)00207-G] [PMID: 7575645]

[42]   Shimizu, K.; Yoshihara, E.; Takahashi, M.; Gotoh, K.; Orita, S.; Urakawa, N.; Nakajyo, S. Mechanism of relaxant response to papaverine on the smooth muscle of non-pregnant rat uterus. *J. Smooth Muscle Res.,* **2000**, *36*(3), 83-91.
[http://dx.doi.org/10.1540/jsmr.36.83] [PMID: 11086880]

[43]   Reynolds, C.D.; Palmer, R.A.; Gorinsky, B. Crystal and molecular structure of the alkaloid papaverine hydrochloride. *J. Cryst. Mol. Struct.,* **1974**, *4*, 213-225.
[http://dx.doi.org/10.1007/BF01198178]

[44]   Peng, Z.; Song, W.; Han, F.; Chen, H.; Zhu, M.; Chen, Y. Chromatographic tandam mass spectrometric detection of papaverine and its major metabolites in rat urine. *Int. J. Mass Spectrom.,* **2007**, *266*, 114-121.
[http://dx.doi.org/10.1016/j.ijms.2007.07.013]

[45]   Hoskin, P.J.; Hanks, G.W. Opioid agonist-antagonist drugs in acute and chronic pain states. *Drugs,* **1991**, *41*(3), 326-344.
[http://dx.doi.org/10.2165/00003495-199141030-00002] [PMID: 1711441]

[46]   https://www.pharmgkb.org/pathway/PA146123006

[47]   Reisfield, G.M.; Salazar, E.; Bertholf, R.L. Rational use and interpretation of urine drug testing in

chronic opioid therapy. *Ann. Clin. Lab. Sci.,* **2007**, *37*(4), 301-314.
[PMID: 18000286]

[48] Bovill, J.G. Mechanisms of actions of opioids and non-steroidal anti-inflammatory drugs. *Eur. J. Anaesthesiol. Suppl.,* **1997**, *15*, 9-15.
[http://dx.doi.org/10.1097/00003643-199705001-00003] [PMID: 9202932]

[49] Ma, G.; Bavadekar, S.A.; Davis, Y.M.; Lalchandani, S.G.; Nagmani, R.; Schaneberg, B.T.; Khan, I.A.; Feller, D.R. Pharmacological effects of ephedrine alkaloids on human α(1)- and α(2)-adrenergic receptor subtypes. *J. Pharmacol. Exp. Ther.,* **2007**, *322*(1), 214-221.
[http://dx.doi.org/10.1124/jpet.107.120709] [PMID: 17405867]

[50] Tseng, Y.L.; Shieh, M.H.; Kuo, F.H. Metabolites of ephedrines in human urine after administration of a single therapeutic dose. *Forensic Sci. Int.,* **2006**, *157*(2-3), 149-155.
[http://dx.doi.org/10.1016/j.forsciint.2005.04.008] [PMID: 15885945]

[51] Garattini, S.; Lamesta, L.; Mortari, A.; Valzelli, L. Pharmacological and biochemical effects of some reserpine derivatives. *J. Pharm. Pharmacol.,* **1961**, *13*, 548-553.
[http://dx.doi.org/10.1111/j.2042-7158.1961.tb11868.x] [PMID: 13703433]

[52] Dasgupta, S.R.; Haley, T.J. Intraventricular administration of reserpine and its metabolites to conscious cats. *Br. J. Pharmacol. Chemother.,* **1957**, *12*(4), 529-531.
[http://dx.doi.org/10.1111/j.1476-5381.1957.tb00177.x] [PMID: 13489186]

[53] Dill, W.A.; Glazko, A.J.; Kazenko, A.; Wolf, L.M. Studies on the metabolism of reserpine. *J. Pharmacol. Exp. Ther.,* **1956**, *118*(4), 377-387.
[PMID: 13385798]

[54] De Costa, C. St Anthony's fire and living ligatures: a short history of ergometrine. *Lancet,* **2002**, *359*(9319), 1768-1770.
[http://dx.doi.org/10.1016/S0140-6736(02)08658-0] [PMID: 12049883]

[55] https://www.drugbank.ca/drugs/DB00696

[56] Duringer, J.M.; Lewis, R.; Kuehn, L.; Fleischmann, T.; Craig, A.M. Growth and hepatic in vitro metabolism of ergotamine in mice divergently selected for response to endophyte toxicity. *Xenobiotica,* **2005**, *35*(6), 531-548.
[http://dx.doi.org/10.1080/00498250500153838] [PMID: 16192106]

[57] Moubarak, A.S.; Rosenkrans, C.F., Jr Hepatic metabolism of ergot alkaloids in beef cattle by cytochrome P450. *Biochem. Biophys. Res. Commun.,* **2000**, *274*(3), 746-749.
[http://dx.doi.org/10.1006/bbrc.2000.3210] [PMID: 10924348]

[58] Dellafiora, L.; Dall'Asta, C.; Cozzini, P. Ergot alkaloids: From witchcraft till *in silico* analysis. Multi-receptor analysis of ergotamine metabolites. *Toxicol. Rep.,* **2015**, *2*, 535-545.
[http://dx.doi.org/10.1016/j.toxrep.2015.03.005] [PMID: 28962389]

[59] http://www.hnkeyuan.com/sports_nutrition/yohimbine-bark.html

[60] Le Verge, R.; Le Corre, P.; Chevanne, F.; Döe De Maindreville, M.; Royer, D.; Levy, J. Determination of yohimbine and its two hydroxylated metabolites in humans by high-performance liquid chromatography and mass spectral analysis. *J. Chromatogr. A,* **1992**, *574*(2), 283-292.
[http://dx.doi.org/10.1016/0378-4347(92)80041-N] [PMID: 1618961]

[61] Le Corre, P.; Dollo, G.; Chevanne, F.; Le Verge, R. Biopharmaceutics and metabolism of yohimbine in humans. *Eur. J. Pharm. Sci.,* **1999**, *9*(1), 79-84.
[http://dx.doi.org/10.1016/S0928-0987(99)00046-9] [PMID: 10494000]

[62] Duflos, A.; Redoules, F.; Fahy, J.; Jacquesy, J.C.; Jouannetaud, M.P. Hydroxylation of yohimbine in superacidic media: one-step access to human metabolites 10 and 11-hydroxyyohimbine. *J. Nat. Prod.,* **2001**, *64*(2), 193-195.
[http://dx.doi.org/10.1021/np000425z] [PMID: 11429998]

[63] Bharucha, A.E.; Skaar, T.; Andrews, C.N.; Camilleri, M.; Philips, S.; Seide, B.; Burton, D.; Baxter, K.; Zinsmeister, A.R. Relationship of cytochrome P450 pharmacogenetics to the effects of yohimbine on gastrointestinal transit and catecholamines in healthy subjects. *Neurogastroenterol. Motil.*, **2008**, *20*(8), 891-899.
[http://dx.doi.org/10.1111/j.1365-2982.2008.01124.x] [PMID: 18433425]

[64] Moudi, M.; Go, R.; Yien, C.Y.; Nazre, M. Vinca alkaloids. *Int. J. Prev. Med.*, **2013**, *4*(11), 1231-1235.
[PMID: 24404355]

[65] van Tellingen, O.; Beijnen, J.H.; Nooijen, W.J.; Bult, A. Tissue disposition, excretion and metabolism of vinblastine in mice as determined by high-performance liquid chromatography. *Cancer Chemother. Pharmacol.*, **1993**, *32*(4), 286-292.
[http://dx.doi.org/10.1007/BF00686174] [PMID: 8324870]

[66] Zhou-Pan, X.R.; Sérée, E.; Zhou, X.J.; Placidi, M.; Maurel, P.; Barra, Y.; Rahmani, R. Involvement of human liver cytochrome P450 3A in vinblastine metabolism: drug interactions. *Cancer Res.*, **1993**, *53*(21), 5121-5126.
[PMID: 8221648]

[67] Dennison, J.B.; Kulanthaivel, P.; Barbuch, R.J.; Renbarger, J.L.; Ehlhardt, W.J.; Hall, S.D. Selective metabolism of vincristine in vitro by CYP3A5. *Drug Metab. Dispos.*, **2006**, *34*(8), 1317-1327.
[http://dx.doi.org/10.1124/dmd.106.009902] [PMID: 16679390]

[68] Dennison, J.B. *Vincristine Metabolism and the Role of CYP3A5*; Indiana University, **2007**.

[69] Dennison, J.B.; Renbarger, J.L.; Walterhouse, D.O.; Jones, D.R.; Hall, S.D. Quantification of vincristine and its major metabolite in human plasma by high-performance liquid chromatography/tandem mass spectrometry. *Ther. Drug Monit.*, **2008**, *30*(3), 357-364.
[http://dx.doi.org/10.1097/FTD.0b013e31816b92c9] [PMID: 18520608]

[70] http://www.news-medical.net/health/Serotonin-in-Plants.aspx

[71] McISAAC, W.M.; Page, I.H. The metabolism of serotonin (5-hydroxytryptamine). *J. Biol. Chem.*, **1959**, *234*(4), 858-864.
[PMID: 13654278]

[72] Rusconi, M.; Conti, A. Theobroma cacao L., the Food of the Gods: a scientific approach beyond myths and claims. *Pharmacol. Res.*, **2010**, *61*(1), 5-13.
[http://dx.doi.org/10.1016/j.phrs.2009.08.008] [PMID: 19735732]

[73] Eteng, M.U.; Eyong, E.U.; Akpanyung, E.O.; Agiang, M.A.; Aremu, C.Y. Recent advances in caffeine and theobromine toxicities: a review. *Plant Foods Hum. Nutr.*, **1997**, *51*(3), 231-243.
[http://dx.doi.org/10.1023/A:1007976831684] [PMID: 9629863]

[74] Weimann, A.; Sabroe, M.; Poulsen, H.E. Measurement of caffeine and five of the major metabolites in urine by high-performance liquid chromatography/tandem mass spectrometry. *J. Mass Spectrom.*, **2005**, *40*(3), 307-316.
[http://dx.doi.org/10.1002/jms.785] [PMID: 15685651]

[75] Yu, C.L.; Louie, T.M.; Summers, R.; Kale, Y.; Gopishetty, S.; Subramanian, M. Two distinct pathways for metabolism of theophylline and caffeine are coexpressed in Pseudomonas putida CBB5. *J. Bacteriol.*, **2009**, *191*(14), 4624-4632.
[http://dx.doi.org/10.1128/JB.00409-09] [PMID: 19447909]

[76] Santos, A.P.; Moreno, P.R. Pilocarpus spp.: A survey of its chemical constituents and biological activities. *Rev. Bras. Cienc. Farm.*, **2004**, *40*, 116-1137.
[http://dx.doi.org/10.1590/S1516-93322004000200002]

[77] Hioki, T.; Fukami, T.; Nakajima, M.; Yokoi, T. Human paraoxonase 1 is the enzyme responsible for pilocarpine hydrolysis. *Drug Metab. Dispos.*, **2011**, *39*(8), 1345-1352.
[http://dx.doi.org/10.1124/dmd.111.038141] [PMID: 21521796]

[78]   Endo, T.; Ban, M.; Hirata, K.; Yamamoto, A.; Hara, Y.; Momose, Y. Involvement of CYP2A6 in the formation of a novel metabolite, 3-hydroxypilocarpine, from pilocarpine in human liver microsomes. *Drug Metab. Dispos.,* **2007**, *35*(3), 476-483.
       [http://dx.doi.org/10.1124/dmd.106.013425] [PMID: 17178767]

[79]   Reyes-Escogido, Mde.L.; Gonzalez-Mondragon, E.G.; Vazquez-Tzompantzi, E. Chemical and pharmacological aspects of capsaicin. *Molecules,* **2011**, *16*(2), 1253-1270.
       [http://dx.doi.org/10.3390/molecules16021253] [PMID: 21278678]

[80]   Chanda, S.; Bashir, M.; Babbar, S.; Koganti, A.; Bley, K. In vitro hepatic and skin metabolism of capsaicin. *Drug Metab. Dispos.,* **2008**, *36*(4), 670-675.
       [http://dx.doi.org/10.1124/dmd.107.019240] [PMID: 18180272]

[81]   Reilly, C.A.; Ehlhardt, W.J.; Jackson, D.A.; Kulanthaivel, P.; Mutlib, A.E.; Espina, R.J.; Moody, D.E.; Crouch, D.J.; Yost, G.S. Metabolism of capsaicin by cytochrome P450 produces novel dehydrogenated metabolites and decreases cytotoxicity to lung and liver cells. *Chem. Res. Toxicol.,* **2003**, *16*(3), 336-349.
       [http://dx.doi.org/10.1021/tx025599q] [PMID: 12641434]

[82]   Corbin, K.D.; Zeisel, S.H. Choline metabolism provides novel insights into nonalcoholic fatty liver disease and its progression. *Curr. Opin. Gastroenterol.,* **2012**, *28*(2), 159-165.
       [http://dx.doi.org/10.1097/MOG.0b013e32834e7b4b] [PMID: 22134222]

[83]   http://lpi.oregonstate.edu/mic/other-nutrients/choline

CHAPTER 5

Natural Products in Cancer Chemoprevention and Chemotherapy

Dev Bukhsh Singh[1,*], Manish Kumar Gupta[2] and **Rajesh Kumar Pathak[3]**

[1] *Department of Biotechnology, Institute of Biosciences and Biotechnology, Chhatrapati Shahu Ji Maharaj University, Kanpur-208024, India*

[2] *Department of Biotechnology, Faculty of Science, Veer Bahadur Singh Purvanchal University, Jaunpur-222003, Uttar Pradesh, India*

[3] *School of Agricultural Biotechnology, Punjab Agricultural University, Ludhiana-141004, Punjab, India*

Abstract: Cancer is a very fatal, challenging and complex disease. A large number of people across the globe are suffering from various types of cancer. The bioactive substances isolated from different parts of several herbs and spices have shown their valuable preventive and therapeutic role against different forms of cancer. The recent technological innovations have made it possible to explore the molecular targets of these bioactive substances and also enabled us to know the mechanism of action related to disease modulation. Our traditional knowledge related to chemopreventive and chemotherapeutic role of herbs is now being validated and explored by the use of modern biological techniques. However, mechanism of action that governs anticancer effect has not been well elucidated for many herbs and natural products. Herbal extracts are the mixture of different active substances, therefore screening and pharmacological response of each individual compound should be validated separately to gain some insight about mechanism of anticancer property. Polyphenols play significant role in initiation, promotion and progression of cancers by modulating the enzymes and signal of diverse pathways related to cellular proliferations, differentiation, angiogenesis, apoptosis and metastasis. These natural bioactives also serve as lead compounds for drug designing and their biological activity can be further optimize by some desired chemical modification. Optimisation of the therapeutic inhibitors can be enhanced using systems biology modelling of the molecular pathways related to the disease. The aim of this chapter is to provide the insight into the molecular basis of preventive and therapeutic effect of natural bioactive substance against cancer diseases.

Keywords: Apoptosis, Cancer, Chemotherapy, Combination therapy, Metastasis, Natural products, Polyphenols Cell division, Telomerase, Topoisomerase, Transcription factor.

---

* **Corresponding author Dev Bukhsh Singh:** Department of Biotechnology, Institute of Biosciences and Biotechnology, Chhatrapati Shahu Ji Maharaj University, Kanpur-208024, India; E-mail: answer.dev@gmail.com

Atta-ur-Rahman (Ed.)

INTRODUCTION

Cancer is a very challenging disease for researchers to find an effective approach for the detection and treatment. Cancer is the result of genetic changes that transform the normally dividing cells into malignant cells. It is an uncontrolled division of cells, where cells divides rapidly without reaching to stage of maturity. There are different types of cancer based on the parts of body and mechanism of disease. Cancer is the leading cause of premature death and physical disability world-wide and adversely affects the social and economic status of a country [1]. General causes of cancer are toxic chemicals, radiations, pathogens and genetic factors. Sign and symptoms of cancers are specific to the parts of body where it develops. Many cancers are identified by the name of body parts where abnormal growth of cell or tissues occurs such as breast cancer, colon cancer, lung cancer, liver cancer, prostate cancer, thyroid cancer and pancreatic cancer. Some of the general signs and symptoms are weight loss, pain, bleeding, fever, persistent cough and unusual tissue masses.

Despite of having advancements in screening, detection and therapeutic approaches for the treatment of cancer, the burden of cancer is expected to increase in future. Major cause of cancer death is the lack of detection in early stage. Treatment strategies for cancer depends on the type and stage of cancer. Natural products have proven their effective role in the prevention of different types of cancer. Natural products have also shown their potential role in treatment of cancer [2]. Modern techniques of biology have played very important role in elucidating the mechanism of action behind the preventive and therapeutic role of herbal products. With the discovery of mechanism that are involved in progression of a cancer, designing of a therapeutic molecule targeting the enzyme can be initiated. Surgery, chemotherapy and radiation therapy are the commonly used approaches for the treatment of cancer.

Natural products have shown their potential role against various types of cancers, such as pancreatic, prostate, skin, gastric, lung, oral, blood, colorectal, liver, head and neck, cervical and breast cancers. Natural compounds may have bioavailability, efficacy, specificity, metabolism, and toxicity related issues when used as a drug. These therapeutic issues may be overcome by performing a series of necessary chemical changes in the lead compounds. A number of natural products have shown their therapeutic role against cancer in preclinical and clinical studies.

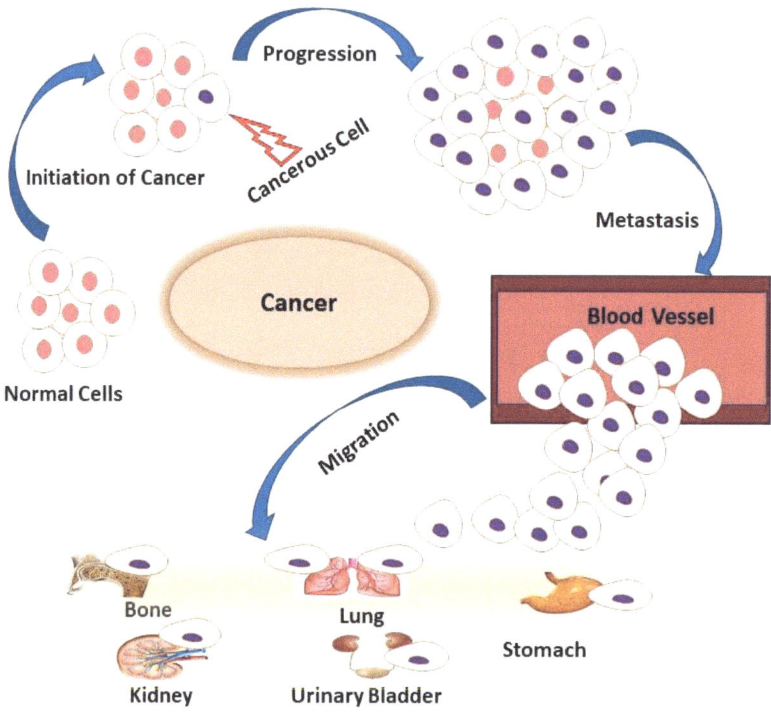

Fig. (1). Initiation, progression and metastasis of cancer.

Cancer is a uncontrolled division of cell due to the accumulation of defects *via* different mechanism, or mutations in DNA and the cells possess the capability to move from one part of body to another Fig. (**1**). Mutations or chromosomal aberrations affecting oncogenes and tumor suppressor genes causes malignant transformation of cells [3]. The most common sites of cancer are breast, cervix lungs, prostate, colorectal, stomach, and liver. Cancer cells can initiate, spread, and grow in various parts of body. The risk of developing cancer increases with age. Altering a diet that includes beneficial phytochemicals can have preventive and therapeutic role against cancer. In cancer chemoprevention, foods containing bioactive chemicals that have anticancer effect can be supplemented in diets. Alkaloids, flavonoids, terpenoids, polysaccharides, saponins, polyphenols and others have been reported as natural products with potential anticancer role [4]. Most of anticancer drugs that are in clinical use for cancer therapy originate from natural products derived from plants, marine sources, and microorganisms. Some natural products have shown their anticancer potential *via* regulating immune function, inducing apoptosis or autophagy, or inhibiting cell proliferation. Other cellular functions that these agents affect include inflammation, anti-oxidant potential, angiogenesis, tissue invasion and cell differentiation.

MECHANISM OF CANCER

Cellular basis of carcinogenesis depends on many genetic changes which finally results in loss of control on cell division. These genetic alternations involves activation of oncogenes, failure of DNA repair mechanism, deregulation of tumour suppression genes. Different mechanistic routes leading to cancer are shown in Fig. (2).

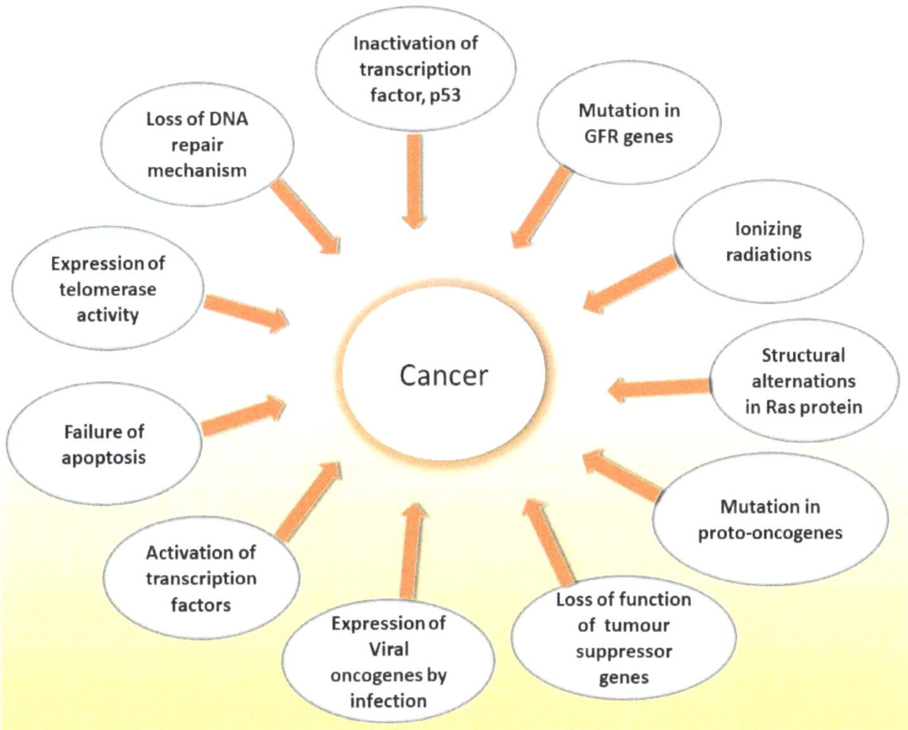

Fig. (2). Different mechanistic routes of cancer.

In normal cell, cell cycle is regulated by a number of proteins that coordinate with each other and ensures the correct completion of different stages of cell cycle. Cyclin dependent kinases (CDKs) play very important role in progression through different phases by phosphorylation. CDK activity depends on the activating subunit called cyclins and forms cyclin-CDK complexes. Cyclin-CDK complexes are regulated by CDK inhibitors. Regulatory proteins such as transcription factor (E2F) activates the S-phase CDKs for starting the DNA synthesis [5]. In normal cell, activation of transcription factor, p53 can arrest cell cycle and induce apoptosis. In most of human cancers the p53 pathway is aberrant. Inactivation of p53 causes failure of apoptosis which finally results in accumulation of cancer cells.

Growth factors play very important role in the normal process of growth by transmitting signal from one to another cells. These signal are received by growth factor receptors (GFRs) and transferred to target molecule that promotes proliferation. For example, activation of signal-transducing protein, Ras activates the MAP kinase. The MAP kinase activates the expression of Myc genes, and transcription factor Myc activates the expression of cell cycles regulatory genes [6]. In this way, the cell enters into cell cycle and results in cell division. Mutation in GFR genes or structural alternations in Ras protein can result in cancer. In most of the cancer, MAP kinase pathways is found deregulated.

Proto-oncogenes regulates the process of cell proliferation and differentiation and they are activated into oncogenes by the point mutations [7]. The products of oncogenes resemble the products of proto-oncogenes but lacks some regulatory elements. Activation of oncogenes may result in deregulation of signals, growth factors, transcription factors and other intermediate components of the MAP kinase pathways.

Loss of function of tumour suppressor genes also results in transformation of a normal cell into cancer cell. Tumour suppressor genes encodes for the proteins that arrest the abnormal cell cycle, negatively regulates the cell cycle and growth signalling pathways, and also promotes the apoptosis and DNA repair mechanism. Examples; tumour suppressor genes adenomatous polyposis coli (APC) suppress the colon cancer by negatively regulating the signalling pathways [8], BRCA1 and BRCA2 genes suppress the breast cancer by promoting DNA repair mechanism [9]. Some viruses are able to induce cancer by integrating and expressing the viral oncogenes in the human. The product of viral oncogenes causes cancer either by interfering with tumour suppressor proteins or by delaying the apoptosis process.

Normal cells possess a limited proliferation capability after that they undergoes senescence. After a certain number of divisions, cells stop dividing due to short length of telomeres. Telomerase is an enzyme which plays very important role in maintaining the length of telomeres [10]. In normal cells, telomerase activity is very slow or zero. In most of cancer cells, telomerase activity is found 10-20 folds higher than normal cells. Suppression of telomerase expression or blocking telomerase activity by inhibitors can be one of the therapeutic approach for the treatment of cancer.

ROLE OF NATURAL PRODUCTS IN CANCER

Natural products are the main source for the discovery and development of anticancer drug. Most of the anticancer drugs are either natural products or inspired or derived from the natural sources. Novel compounds from the natural sources can be identified and developed as safe and potential drug for the cancer therapy. Recent developments and advance in the field of secondary metabolite production, combinatorial chemistry, high-throughput screening and computational chemistry approaches have speed up the process of drug development [11]. A plethora of naturally occurring compounds have been screened, identified and tested in clinical trials and their potential for anticancer therapy have also been validated [12]. These natural compounds generate their preventive and therapeutic role against cancer by targeting various pathways related to cell cycle, cell division and signalling molecules. Bioactive natural products are widely used in clinical practices due to their safety profile.

Natural products have shown their significant role in cancer chemoprevention and chemotherapy (Table **1**). Nature can provide us a wide range of chemical diversity for anticancer therapy. Optimization of natural lead compounds into anticancer drug improves drug efficacy, absorption, distribution, metabolism, excretion and toxicity profiles. Optimization process involves a broad range of chemical changes such as aryl, alkyl or ring substitution, chain or ring expansion or contraction, ring variation, extension of extra function group, simplification, rigidification, blocking the active functional groups, use of bioisosteres and removal of toxic substructure [13].

Some natural compounds and their metabolites can act as carcinogen leading to alternation of normal cellular mechanism [14]. Herbal medicine can also have their side effects due to presence of contaminations. Herbal medicines may also contain non-organic pollutants such as arsenic, lead and mercury which can result in some side effects. Aristolochia, a Chinese herb contains a strong carcinogen aristolochic acid which causes renal failure and also associated with urothelial carcinoma [15]. Anticancer effect of aristolochic acid has been reported due to inhibition of phospholipase A 2 [16]. Sometimes, a single herb may be dangerous for a patient but the same herb in combination with other herbs can generate a synergistic effect. Pharmacodynamics interactions influence qualitative activities of medicine at different level. Herb-herb interactions can generate synergistic effect or can result in antagonizing property. Herbal compounds can affect ADMET of administered drug. Such interactions are clinically important, especially when herbal compounds affect liver enzymes [17].

Table 1. Natural compounds their source and target/mechanism for chemoprevention and chemotherapy.

S. No.	Compound Name	Source	Target/Mechanism	References
1.	Aristolochic acid	Aristolochia	Inhibition of phospholipase A 2	[15, 16]
2.	Curcumin	Turmeric	Inhibition of NF-kB	[18]
3.	Celastrol	*Tripterygium wilfordii*	Inhibition of NF-kB	[19]
4.	Oleanolic acid	Diverse plants (*Hyptis capitata*)	Modulation of transcriptional factor	[20]
5.	Ursolic acid	Diverse fruits and herbs (Apple)	Modulation of transcriptional factor	[20]
6.	Boldine	boldo tree	Modulation of transcriptional factor	[20]
7.	Resveratrol	Diverse plants (*Polygonum cuspidatum*)	Modulation of transcriptional factor	[20]
8.	Genistein	Diverse plants (legumes, soybeans)	Modulation of transcriptional factor	[20]
9.	Emodin	*Cascara sagrada, Polygonum cuspidatum*	Modulation of transcriptional factor	[20]
10.	Dentatin	*Clausena excavata*	Modulation of transcriptional factor	[20]
11.	EGCG	*Camellia sinensis*	Modulation of transcriptional factor	[20]
12.	Garcinol	*Garcinia indica*	Modulation of transcriptional factor	[20]
13.	Nordihydroguaiaretic acid	*Larrea tridentata*	Inhibition of lipoxygenases	[21]
14.	baicalein	*Scutellaria baicalensis, Scutellaria lateriflora*	Inhibition of lipoxygenases	[21]
15.	Rutin	*Ginkgo biloba*	Inhibition of phospholipases A2	[22]
16.	Asiaticoside	*Centella asiatica*	Inhibition of phospholipases A2	[23]
17.	Piperine	*Piper sp.*	Inhibition of phospholipases A2	[21]
18.	Quercetin	*Azardirecta indica*	Inhibition of phospholipases A2	[21]
19.	Coumestans	*Eclipta alba*	Inhibition of phospholipases A2	[21]

(Table 1) cont.....

S. No.	Compound Name	Source	Target/Mechanism	References
20.	Curcumin	*Curcuma longa*	Inhibition of estrogen receptor signaling	[24]
21.	Caffeine	Diverse plants (Coffee and tea)	Inhibition of estrogen receptor signaling	[24]
22.	Coumarin	Cassia leaf, lavender and cinnamon bark	Inhibition of estrogen receptor signaling	[24]
23.	Ellagic acid	Diverse plants (strawberries, grapes)	Inhibition of estrogen receptor signaling	[24]
24.	EGCG	*Camellia sinensis* (green tea)	Inhibition of estrogen receptor signaling	[24]
25.	Luteolin	Diverse plants (green peppers)	Inhibition of estrogen receptor signaling	[24]
26.	Lycopene	Tomato	Inhibition of estrogen receptor signaling	[24]
27.	Morin	*Maclura pomifera*	Inhibition of estrogen receptor signaling	[24]
28.	Myricetin	Diverse plants (grapes, berries)	Inhibition of estrogen receptor signaling	[24]
29.	Quercetin	Diverse plants (onions, green tea, apples)	Inhibition of estrogen receptor signaling	[24]
30.	Resveratrol	Diverse plants (*Polygonum cuspidatum*)	Inhibition of estrogen receptor signaling	[24]
31.	Sesamin	*Sesamum indicum*	Inhibition of estrogen receptor signaling	[24]
32.	Tangeretin	Tangerine and other citrus peels	Inhibition of estrogen receptor signaling	[24]
33.	allicin	Garlic	Inhibition of telomerase	[25]
34.	curcumin	Turmeric	Inhibition of telomerase	[25]
35.	silbinin	Cruciferous vegetables	Inhibition of telomerase	[25]
36.	EGCG	Green tea	Inhibition of telomerase	[25]
37.	sulforaphane	Broccoli and cauliflower	Inhibition of telomerase	[26]
38.	Curcumin	*Curcuma longa*	Promoting apoptosis	[27]
39.	Biocalein	Shosiko	Promoting apoptosis	[27]
40.	Resveratrol	Diverse plants (*Polygonum cuspidatum*)	Promoting apoptosis	[27]
41.	Genistein	Diverse plants (legumes, soybeans)	Promoting apoptosis	[27]

(Table 1) cont.....

S. No.	Compound Name	Source	Target/Mechanism	References
42.	7-hydroxystaurosporine (UCN-01), staurosporine	*Streptomyces staurosporeus*	Promoting apoptosis	[27]
43.	Ginger	*Zingiber officinale*	Promoting apoptosis	[27]
44.	Quercetin	Diverse plants (onions, green tea, apples)	Apoptosis /caspase9	[27]
45.	Magnolal	*Magnolia officinalis*	Promoting autophagy	[28]
46.	Apigenin	Diverse plant (celery, parsley)	Promoting autophagy	[28]
47.	Neoalbaconol	*Albatrellus confluens*	Promoting autophagy	[28]
48.	Salvianolic acid B	*Salvia miltiorrhiza*	Promoting autophagy	[28]
49.	Honokiol	*Magnolia spp.*	Promoting autophagy	[28]
50.	Kazinol A	*Broussonetia papyrifera*	Promoting autophagy	[28]
51.	Alisol B	*Rhizoma alismatis*	Promoting autophagy	[28]
52.	Capsaicin	chili peppers	Promoting autophagy	[28]
53.	Juglanin	*Polygonum aviculare*	Promoting autophagy	[28]
54.	Oridonin	*Rabdosia rubescens*	Promoting autophagy	[28]
55.	Salvigenin	*Salvia officinalis*	Promoting autophagy	[28]
56.	Silibinin	*Silybium marianum*	Inhibition of EGF ligand-induced expression	[29]
57.	Honokiol	Magnolia officinalis	Inhibition of AKT and STAT3 growth inhibition	[29]
58.	Houttuyninum	*Houttuynia cordata*	Inhibition of HER2 phosphorylation	[29]
59.	Taspine	Magnolia spec. Croton lechleri	Inhibition of EGFR	[29]
60.	EGCG	*Camelia sinensis*	Inhibition of HER2 phosphorylation	[29]
61.	Curcumin	*Curcuma longa*	Inhibition of EGFR, HER2, HER3,	[29]
62.	Shikonin	*Lithospermum erythrorhizon*	Inhibition EGFR phosphorylation	[29]
63.	Resveratrol	Diverse plants	Inhibition of EGF-mediated cell migration	[29]
64.	Vinblastine	*Fusarium oxysporum*	Interfering with microtubules	[30]

(Table 1) cont.....

S. No.	Compound Name	Source	Target/Mechanism	References
65.	Vincristine	*Fusarium oxysporum*	Interfering with microtubules	[30]
66.	Taxane	*Taxus breviflora*	Interfering with microtubules	[30]
67.	Noscapine	Opium (poppy)	Interfering with microtubules	[30]
68.	Colchicines	*Colchicum autumnale L.*	Interfering with microtubules	[30]
69.	Lamellarin D	Lamellaria	Inhibition of topoisomerase	[31]
70.	Camptothecins	*Camptotheca acuminata*	Intercalates between the DNA base pairs at the topoisomerase cleavage site	[32]
71.	Erybraedin C	*Bituminaria bituminosa*	Inhibition of topoisomerase	[31]
72.	Thaspine	*Croton lechleri*	Inhibition of topoisomerase	[31]
73.	Evodiamine	*Evodia rutaecarpa*	Inhibition of topoisomerase	[31]
74.	Albanol A	*Morus alba*	Inhibition of topoisomerase	[31]
75.	Alternariol	*Alternaria alternata*	Inhibition of topoisomerase	[31]
76.	4-hydroxyderricin	*Angelica keiskei*	Inhibition of topoisomerase	[31]
77.	Echinoside A	*Actinopyga echinites*	Inhibition of topoisomerase	[31]
78.	Riccardin D	Chinese liverwort	Inhibition of topoisomerase	[31]
79.	Wedelolactone	*Eclipta alba*	Inhibition of topoisomerase	[31]
80.	Tricitrinol B	*Penicillium citrinum*	Inhibition of topoisomerase	[31]
81.	Daurinol	*Haplophyllum dauricum*	Inhibition of topoisomerase	[31]
82.	Fisetin	Diverse plants (strawberries, apples)	Inhibition of topoisomcrase	[31]
83.	Myricetin	Diverse plants (grapes, berries, fruits)	Inhibition of topoisomerase	[31]

(Table 1) cont.....

S. No.	Compound Name	Source	Target/Mechanism	References
84.	Sulfonoquinovosyl diacylglyceride	*Azadirachta indica*	Inhibition of topoisomerase	[31]
85.	Plumbagin	*Plumbago zeylanica*	Inhibition of topoisomerase	[31]
86.	Bryostatin-1	*Bugula neritina*	Activates protein kinase C (PKC) isoenzymes	[32]
87.	Aplysiatoxin (ATX)	cyanobacteria species	Activates PKC isoenzymes	[33]
88.	(3R)-icos-(4E)-en-1-yn-3-ol	*Cribrochalina vasculum* (sponge)	Inhibitor of insulin-like growth factor-1 receptor (IGF-1R)	[34]
89.	Hymenialdisine and debromohymenialdisine	*Stylotella aurantium* (sponge)	Inhibitor of cyclin-dependent kinases (CDKs)	[35]
90.	Fascaplysin	*Thorectandra* (marine sponge)	Selective inhibitor of CDK4	[36]
91.	Meridianins A–G	*Aplidium meridianum*	Inhibit CDK1, and CDK5 (ATP binding site)	[37]
92.	palinurin		Glycogen synthase kinase-3 beta (GSK-3β)	[38]
93.	Manzamine A	*Acanthostrongylophora*	non-competitive inhibitor of ATP with binding to GSK-3β	[39]
94.	Phenylmethylene hydantoin	*Hemimycale arabica*	Inhibitor of GSK-3β	[40]
95.	pannorin	Penicillium sp.	Inhibitor of GSK-3β	[41]
96.	Anthraquinone derivatives 1'-deoxyrhodoptilometrin	*Comanthus* sp. (Echinoderm)	inhibitors of IGF-1R, and EGFR	[42]
97.	BDDPM	*Rhodomelaceae confervoides*	Receptor tyrosine kinase (RTK) inhibitor	[43]
98.	Pyrroloiminoquinone	*Latrunculia* sp.,	HIF-1α/p300 inhibitors	[44]
99.	Psammaplin A	*Pseudoceratina purpurea*	inhibitor of Histone deacetylase (HDAC)	[45]
100.	Largazole	*Symploca* sp (cyanobacterium)	HDAC inhibitor	[46]
101.	epipolythiopiperazine-2, 5-diones	*Oidiodendron truncatum* *(*fungus*)*	Hsp90 inhibitor	[47]

(Table 1) cont.....

S. No.	Compound Name	Source	Target/Mechanism	References
102.	12β-(3′β-hydroxybutanoyloxy)-20	*Carteriospongia* sp., (sponge)	Hsp90 and topoisomerase II inhibitor	[48]
103.	Spongiacidin C	*Stylissa massa* (sponge)	Ubiquitin specific peptidase 7 (USP7)	[49]
104.	Sipholane triterpenoids	*Callyspongia siphonella*	Inhibitor of P-glycoprotein (P-gp)	[50]
105.	Swinhosterol B	*Theonella swinhoei*	pregnane X receptor (PXR) agonist and	[51]

Cancer Chemoprevention

Several clinical trials have been conducted since 2006 to evaluate the enhancing effect of natural compounds such as curcumin or traditional Chinese medicine (TCM) in the promotion of conventional chemotherapy against different cancers [52]. Based on the study, it was suggested that the natural agents are working as adjuvants in various cancers. Besides, herbal medicinal products such as astragalus [53] and turmeric [54] have also been reported as potential cancer adjuvants [55]. Natural compounds used in folk and traditional medicine have been recognized as one of the main sources for the discovery and development of chemopreventive drugs [56]. Carcinogenesis is a multi-stage process that transforms a normal cell into a cancer cell. Chemopreventive molecules are able to target different stages of cancer development and increasing apoptosis, altering gene expression and decreasing angiogenesis [56]. Natural compounds are reported to affect multiple targets with lower side effects and to be effective against multiple types of cancer. Therefore, it is used is cell culture or in clinic study for exploring its inhibitory potential, efficacy and potency for further investigation and development of chemopreventive agent [57].

Chemopreventive agents isothiocyanates naturally found in cruciferous vegetables such as phenethyl isothiocyanate, sulforaphane and benzyl isothiocyanate are effective inhibitors of tumor initiation in carcinogens treated rodents [58]. It is having ability to modifying carcinogen metabolism through Phase 1 enzymes inhibition and/or Phase 2 enzymes induction. These effects are highly precise and unique, because they are depends on isothiocyanate and carcinogen structure. Phenethyl isothiocyanate inhibition of 4-(methylnitrosamino)-1-(3-pyrid-1)-1-butanone (NNK)-induced lung tumorigenesis is one of the most widely studied examples of isothiocyanate inhibition of rat carcinogenesis [58]. Besides, several chemopreventive and dietary agents such as Chlorophyll, chlorophyllin and dithiolethiones are reported to inhibit DNA adduct formation and helped in

DNA repair mechanism [59 - 64]. The role of chemopreventive agents includes inhibition of carcinogenic activation, scavenging of free radicals and reactive carcinogenic metabolites, as well as increased carcinogen detoxification by modulating cellular metabolism [65].

Ayurveda, India's ancient Vedic literature is the science of health and well-being, and collection of traditional plant based knowledge for curing diseases. In the current healthcare system, modern drug development program based on ayurvedic concepts has gained wide acceptance [66]. Besides, a big number of scientific literatures are available, highlighting the role of food based compounds in prevention and cure of different diseases including cancer *i.e.* genistein, diadzei from soybean; resveratrol from grapes; lycopene from tomatoes, EGCG, EGC and ECG from green tea; silibinin and silychristin from milk thistle; glucoraphan and sulforaphane from broccoli; punicalagin from pomegranate; anthocyanins, ellagic acid and urolithin A from black raspberry; curcumin from turmeric and many more found in different parts of plants [67 - 69]. Therefore, chemoprevention agents that are food-based such as broccoli sprouts, freeze-dried berries, green and black tea, tomato paste, extracts from several foods such as berries, beets, and other sources [70].

In early 2008, 225 natural product-based drugs were used in different testing methods, such as preclinical, phases I to III clinical trials and pre-registration. Of these, 108 were plant-based, 61 were semi-synthetic, 25 were bacterial, 24 were animal, and 7 were originated from fungal sources [71]. Besides, clinical trials were also conducted on 22 compounds out of 18000 known marine natural products, or their chemical derivatives [71]. However, other plant natural products such as the (−)-epigallocatechin-3-gallate, isoflavone genistine, curcumin, indole-3-carbinol (I3C), 3,3′-diindolemethane, resveratrol and lycopene, are known to inhibit cancer cell development [72]. Quarfloxin, reached phase II in clinical trials for the treatment of cancer. However, its limited bioavailability has prevented further progress [73, 74].

Cancer Chemotherapy

Over the past years, natural products have been become the pillar of cancer chemotherapy [75]. More than 60% of the existing anti-cancer medicines are derived from natural sources in one way or another [76]. Historically, plants have been the main source of natural drug discovery progam, and plant-derived agents such as VBL and vincristine (VCR), paclitaxel (taxol ®), etoposide, docetaxel, irinotecan and topotecan are among the most effective cancer chemotherapy available today [77]. Besides, significant numbers of marine-derived antitumor agents are investigated showing a potent growth inhibition of human tumor cells *in vitro* and, in many cases, *in vivo* murine models and in humans have been determined, but although many agents have been clinically tested for cancer, only four have been approved for human use to date. These agents are cytarabine, trabectedin, halichondrin B, and eribulin [78].

However, some molecules having chemotherapeutics potential were also investigated in fungi and bacteria. Among the most effective cancer chemotherapeutic agents are anti-tumor antibiotics. These are isolated from the members of actinomycin, anthracycline, ansamycin, epothilone, bleomycin, and classes of staurosporin. The famous microbial-derived chemotherapeutic agents are rapamycins, carfilzomib, and midostaurin [78]. Therefore, natural product research for chemotherapy is an important approach to the discovery of biologically active novel compound with their mode of action for saving human life [79].

The common cancer treatments are chemotherapy and radiotherapy. The occurrence of adverse effects from chemotherapy and radiotherapy, however, hinders therapeutic use and decreases the quality of life of patients with cancer. There is evidence that natural products including crude extracts, bioactive fractions enriched with components and pure herbal compounds can minimize side effects from such therapy *i.e.* oral mucositis, gastrointestinal toxicity, hepatotoxicity, damage to the hematopoietic system, nephrotoxicity, cardiotoxicity, and neurotoxicity [80]. Therefore, natural dietary supplements are able to neutralize side effects of chemotherapy and radiotherapy. Besides, natural polymers from plants, animals and microbes are commonly used in cancer vaccines. They have excellent antigen delivery capacities, APC activation and cross-presentation antigen induction based on their physicochemical properties [81].

THERAPEUTIC MECHANISM OF ACTION

Recognizing the mechanism behind progression of different types of cancer can be one of the remarkable step in treatment of various cancers. Different therapeutic mechanism and targets through which natural compounds exerts their anti-cancer effects are discussed here Fig. (**3**).

Modulation of Diverse Transcription Factors

The transcription factors play very important role in initiation, progression and metastasis of cancer. Activation of different transcription factors induces the cells to enter in cell cycle and finally results in cancer. One of the major problem associated with cancer treatment of cancer is the activation of pro-inflammatory transcription factors such as nuclear factor-κB, signal transducer and activator of transcription 3, activator protein 1, fork head box protein M1, and hypoxia-inducible factor 1α [20]. In case of cancer, the production of cellular adhesion molecules, chemokines, inflammatory cytokines, anti-apoptotic molecules, and inducible nitric oxide synthase increases very rapidly. A number of natural products possess the capability to modulate the activation of diverse oncogenic transcription factors. The combination therapy of drugs with natural compounds has shown a significant effect on survivality of cancer patients. Curcumin inhibits the growth of lung cancer, prostate cancer and breast cancer by inhibiting the pro-inflammatory transcription factor, NF-kB [18]. Curcumin also inhibit Wnt/β-catenin signaling and reversed Wnt/β-catenin-induced cell invasion, migration, and proliferation [82]. Celastrol is isolated from the root extracts of *Tripterygium wilfordii*. Celastrol inhibits hematopoietic cancer cells by negatively regulating the NF-kB signaling pathway [19]. Some other natural compounds such as oleanolic acid, ursolic acid, boldine, resveratrol, genistein. emodin, dentatin, EGCG and garcinol have shown their anti-cancer property by targeting different oncogenic transcription factors [20]. EGCG has shown good binding affinity for cancer drug target AP-1 protein [83]. In-depth knowledge of mechanism of action for these compounds can help in translating the potential of these compounds in the form of an active drug.

Targeting Arachidonic Acid Pathway

Metabolic enzymes of Arachidonic acid pathway, phospholipase A2s, cyclooxygenases and lipoxygenases and their metabolic products, such as prostaglandins and leukotrienes, are used as novel targets in cancer [21]. Curcumin,, anthocyanins, resveratrol, apigenin berberine, eugenol, ellagic acid, fisetin, ursolic acid, guggulsteone, gingerol, lycopene and genistein are well known cancer chemopreventive agents which act by targeting multiple pathways. Nordihydroguaiaretic acid and baicalein are used as chemopreventive molecules against various cancers by inhibiting lipoxygenases. Several natural products against cancer were also identified that causes their effect by inhibiting the phospholipase A2. Phospholipases A2 are a diverse group of enzymes that hydrolyze membrane phospholipids into arachidonic acid and lysophospholipids [84]. Arachidonic acid get metabolised into eicosanoids such as prostaglandins, thromboxanes, leukotrienes), and lysophospholipids are converted to platelet-activating factors. Many synthetic inhibitors (methyl arachidonyl fluorophosphonate; bromoenol lactone; indole-based inhibitors; pyrrolidine-based inhibitors; amide inhibitors, 1,3-disubstituted propan-2-ones and polyfluoroalkyl ketones) and natural product based inhibitors (curcumin, *Ginkgo biloba* and *Centella asiatica* extracts) of Phospholipases A2 have been discovered and used for the treatment of neurological disorders and oxidative stress [22].

Inhibition of Estrogen Receptor Signaling

Estrogen receptor (ER)-positive breast cancer is the most common type of breast cancer. A better understanding of ER-positive breast cancer will help in designing and development of therapeutics for breast cancer. Development of breast cancer depends on genetic factors as well as lifetime exposure to estrogen. The ER-dependent pathway of breast cancer involves cell growth and proliferation triggered by the binding of estrogen to the ER. In ER-independent mechanisms, genotoxic metabolites, free radicals and reactive oxygen species produced as result of estrogen metabolism, induces breast cancer [85]. Many natural products and bioactive compounds have been shown to inhibit progression of breast cancer *via* inhibition of estrogen induced oxidative stress as well as ER signaling. Phytochemicals includes a broad set of chemical compounds such as polyphenols, flavonoids, steroidal saponins, organosulphur compounds and vitamins. A number of bioactive natural products such as curcumin, caffeine, coumarin, ellagic acid, EGCG, luteolin, lycopene, morin, myricetin, quercetin, resveratrol, sesamin and tangeretin have shown their therapeutic role against ER-positive breast cancer *in vivo* [24]. Some natural bioactives such as curcumin, EGCG, flaxseed, genistein, indole-3-carbinol, lycopene, sulforaphane and tocotrienol have completed the different trial phases of clinical studies.

Targeting Telomerase

Telomere length is maintained by the human reverse transcriptase telomerase (hTERT), which synthesizes the new telomeric DNA from a RNA. Higher activity of telomerase is observed in most of the cancer cells. Thus, reversal of telomerase up regulation is a potential strategy for treatment of cancer. Natural products, synthetic inhibitors, vaccines, oligonucleotide inhibitors, and gene therapy approaches can be used for treatment of cancer. The selective inactivation of telomerase in cancer cell can be carried out as most of the somatic cells no or very limited telomerase activity. The natural compounds such as allicin, from garlic; curcumin, from turmeric; silbinin, from cruciferous vegetables; epigallocatechin gallate (EGCG), from green tea, have potential to inhibit the activity of telomerase [25]. The role of curcumin, genistein, EGCG, and sulforaphane was tested in breast cancer cells. The mode of action of these compounds is partially known which involves inhibition of translocation of hTERT to the nucleus; dissociation of Hsp-90 co-chaperone from hTERT; and a decrease of hTERT activity [26]. Studies have shown that lifespan of pancreatic cancer cells can be limited by telomerase inhibition. Imetelstat (GRN163L), an inhibitor of telomerase is under clinical trial [86]. The activity of allicin, curcumin, silbinin, EGCG and other lead compounds can be optimized to achieve potential inhibition of telomerase. There are many natural compounds such as curcumin, allicin, lycopene, EGCG, resveratrol and genistein which possess the capability to prevent and cure the cancer by targeting different mechanistic pathways.

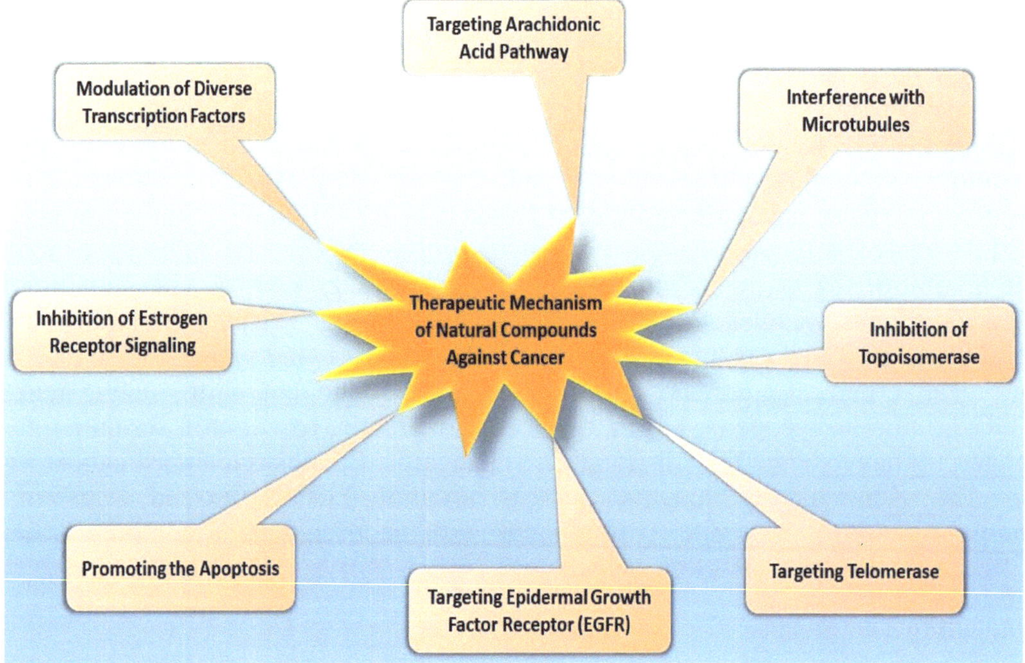

Fig. (3). Therapeutic role of natural compounds against cancer.

Promoting the Apoptosis

Apoptosis or programmed cell death is a well regulated pathway that is responsible for the removal of diseased, injured and old cells. Failure of apoptosis mechanism results in cancer. Several morphological and molecular changes in a cell induces the apoptosis. Apoptosis is either induced by death receptors, caspase 8 and 10 and or by caspase 9 of mitochondrial pathways. Some natural compounds act as anticancer agent by inducing the apoptosis in cancer cells through various mechanism. Curcumin, biocalein, resveratrol, genistein, 7-hydroxystaurosporine (UCN-01), ginger, quercetin and some other natural compounds have shown their potential to induce the apoptosis in different types of cancer cells [27]. Biocalein induces apoptosis in cell lines of human hepato-cellular carcinoma. Resveratrol causes apoptosis in HL60 cells and breast carcinoma cells. Genistein induces apoptosis in human pro-myelocytic HL-60 leukaemic cells. Quercetin induces apoptosis through release of cytochrome-c to the cytosol and activating of caspase9. It is accompanied by Mcl-1 down-regulation and Bax conformational change and mitochondrial translocation which triggered cytochrome-c [87]. Autophagy can relieve tumor cells from nutrient and oxidative stress and maintains the homeostatic process by cell death and tumour shrinkage. Natural products such as curcumin, 16-hydroxycleroda-3,13-d-en-15,16-olide, and prodigiosin are known to induce autophagy that triggers cell death [28]. A number of natural compounds such as magnolal, apigenin, neoalbaconol, salvianolic acid B, honokiol, oridonin, kazinol A, alisol B, capsaicin, juglanin, oridonin and salvigenin induces the autophagy in different types of cancer cells through various pathways [28].

Targeting Receptors (GFRs)

Growth factors play important role in regulating various cellular processes such as cell migration, cell invasion, apoptosis, angiogenesis and survival of cells. Deregulation of expression of GFRs leads to cancer initiation and progression. However, many inhibitors and monoclonal antibodies are in clinical practice but these approaches are still not very safe and effective. Short interfering RNA (S-RNA) based therapies using liposome-based nanocarriers for delivery have shown significant role in suppressing GFRs expression [88]. Tyrosine kinase inhibitors such as gefitinib, erlotinib, axitinib, and sunitinib have been approved for the treatment of different cancers but these are not natural compounds. Phytochemicals such as silibinin, honokiol, taspine, genistein, curcumin, shikonin, resveratrol, and betulinic acid have shown their activity against EGFR and their clinical utility should be tested. Gefitinib binds reversible to the ATP binding site of receptor and inhibits anti-apoptotic signals. Erlotinib induces cell cycle arrest and apoptosis in cancer cells.

Binding of ligand to epidermal growth factor receptors (EGFR) activate the pathways involved in cancer proliferation, differentiation, metastasis and apoptosis The EGFR is a receptor tyrosine kinase that is commonly upregulated in non-small-cell lung cancer, metastatic colorectal cancer, head and neck cancer, glioblastoma, pancreatic cancer, and breast cancer. Upregulation of EGFR activity occurs due to various mechanism which includes mutations and truncations to its extracellular domain (EGFRvIII), L858R and T790M mutations, or the exon 19 truncation [89]. There are four EGFRs HER1, HER2, HER3, and HER4. EGFR is expressed in cancers of the lung, colon, pancreas, breast, ovary, bladder, and kidney. EGFRs are an important molecular target in cancer therapy. Due to development of drug resistance, the monoclonal antibodies and drugs targeting tyrosine kinase receptor have shown toxic and fatal outcome in cancer patients.

Resistance against tyrosine kinase inhibitors have been reported due to point mutation in EGFR variants [90]. Therefore, natural products targeting different GFRs can be a better and safe alternative for chemotherapy of cancers. Phytochemicals such as silibinin, honokiol, taspine, genistein, curcumin, shikonin, resveratrol, and betulinic acid have shown their activity against EGFR and their clinical utility should be tested. Curcumin, EGCG, houttuyninum, and apigenin were found active against HER2 and HER3 [29]. Curcumin causes inhibition of expression and activation of EGFR, HER2, and HER3 in colon cancer. It has shown synergistic growth inhibition with 5-fluorouracil plus oxaliplatin [91]. Resveratrol causes inhibition of EGF-mediated cell migration in MDA-MB-231 breast cancer [92]. EGCG, houttuyninum and genistein exert their anticancer effect by inhibition of HER2 phosphorylation and STAT3 activation in different types of breast cancer cells [29].

Interference with Microtubules

Microtubules, are the major cytoskeletal components in all eukaryotic cells. Microtubules play important role in maintenance of cell shape and polarity, transportation of vesicles and organelles, and action of cilia and flagella [93]. Microtubules also controls the movement of daughter chromosomes from equator to poles. Drugs targeting microtubules stops the mitosis at metaphase and anaphase, and these cells undergoes the process of apoptosis. Some natural compounds such as vinca alkaloids, vinblastine and vincristine, taxane, noscapine, colchicines have shown their anti-cancer effect by targeting microtubules [30]. This study has promoted the discovery of some semi-synthetic drugs such as vindesine, vinorelbine and vinflunine, which are now used as anti-cancer agent. At lower concentration, taxane suppress the movement mechanism of microtubules and prevents the mitosis and subsequent cell death by apoptosis. Noscapine is an alkaloid obtained from *Papaver somniferum*. It also acts as an anti-cancer agent by suppressing microtubule dynamics [94].

Inhibition of Topoisomerase

Topoisomerases are essential for DNA replication, transcription and recombination. The enzymes topoisomerase controls the topology of DNA molecules at different steps in replication, transcription and recombination. Topoisomerase has been reported as one of the attractive drug target against cancer [95]. A number of natural bioactive and derivative have shown their inhibitory role against topoisomerase.

Camptothecins intercalates between the DNA base pairs at the topoisomerase cleavage site. This intercalation induces an accumulation of cleaved DNA, which results in damage of chromosome [96]. Etoposide is a chemotherapeutic agent that targets topoisomerase II. Etoposide provides stability to covalent enzyme-cleaved DNA complex in the catalytic cycle of topoisomerase II [97]. A number of natural topoisomerase inhibitors such as lamellarin D, erybraedin C, thaspine, evodiamine, albanol A, alternariol, 4-hydroxyderricin, echinoside A, riccardin D, wedelolactone, tricitrinol B, daurinol, fisetin, myricetin, sulfonoquinovosyl diacylglyceride and plumbagin have shown their preventive and therapeutic effect against cancer [31]. The clinical use of topoisomerase inhibitors is limited because, they damage DNA of normal cell and also show low potency at normal physiological pH. Broad spectrum of chemical and structural diversity for these inhibitors offer different novel skeletons for development of more effective drugs against cancer.

COMBINATION THERAPY OF NUTRACEUTICAL

Nutraceutical are dietary supplements which provides some health benefits in addition to the basic nutritional value. An effective nutraceutical should have no or low toxic effect with potential health benefits. A combination of bioactive molecule with an effective synthetic drug or another bioactive molecule can generate a synergistic or antagonistic effect against a disease. Dietary composition of the molecule can be adjusted as per the outcome of treatment. Genetic changes, social determinants, population, and dosage variations affects the therapeutic efficacy of nutraceuticals [98].

Many cancer-related deaths happen due to lung and bronchus, breast, colorectal, and prostate cancer. These cancers are less prevalent in Asian countries due to more consumption of vegetables, fruits, spices and other herbs. The use of nutraceutical is safer as compared to synthetic drug because naturally occurring chemicals have minimum side effect. Nutraceutical can be used as a template for combinatorial synthesis to increase the efficacy of a promising nutraceutical. The molecular targets of nutraceuticals should be identified; it can guide us to design more potential compounds that specifically binds with binding site of targets [99]. A combination of taxol with curcumin downregulate the expression of NF-κB and induce apoptosis in cancer cells. A sequential addition of curcumin and xanthorrhizol in culture can show synergistic and antagonistic effects depending on which compound was added first to the culture [100]. This indicates that efficacy of a drug is dependent on dose, time, and how supplemented with other. Docetaxel is used to inhibit the assembly of microtubules in cancerous cells. It also inhibits the genes which regulate cell proliferation, spindle formation, transcription factors, and oncogenesis. D-Limonene has shown therapeutic effect against prostate cancer at lower dose. Therefore, combination of docetaxel and d-Limonene have shown potential therapeutic role against prostate cancer [101]. Limonene and its main metabolites perillyl alcohol have anticancer activity. However, their low metabolic and plasma stability, low bioavailability, and tissue distribution / accumulation could limit their beneficial health properties [102]. Besides, perillyl alcohol is able to inhibit the tumor cells growth in cell culture and enhance preventive and therapeutic activity in different animal tumor models. However, it did not fulfil its promise as a novel anticancer agent during oral administration at the time of clinical trials. The clinical trials conducted in Brazil investigated the delivery of intranasal perillyl alcohol as an alternative to the toxic limitations of oral administration. In this trials, Patients with recurrent malignant gliomas were provides relatively small doses of perillyl alcohol through simple nose inhalation. Results from these studies show that this form of long-term, routine chemotherapy is well tolerated and effective [103].

Individual effect of a compound show higher dose requirements, whilst their combination may generate same effect on lower dose. Synergistic effects of compounds occur if their individual functions are enhanced in the presence of other compounds. Drug or herbal combinations that simultaneously impact multiple targets of disease may be the more effective approach for the treatment of cancer [104]. The mode of delivery of dietary supplements should be considered during trials and studies. Some studies have shown that encapsulation of nutraceutical in nanoparticles can enhance the delivery and increase its stability and bioavailability.

NATURAL PRODUCTS AS A LEAD FOR CANCER THERAPY

Nature provides a vast set of chemical diversity. Novel structural scaffolds also provides a way to overcome drug resistance. Novel natural products serve as lead compounds for drug designing. A plethora of data related to natural products are available in different databases with their complete structural and functional details. Identification of lead compounds from natural resources is not a very complex task. Plant derived anticancer agents in clinical use are vinca alkaloids, taxanes, podophyllotoxin, camptothecins. Vinca alkaloids of vinblastine and vincristine was first herbal anti-cancer medicine clinically used in 1960 [105]. Vinca alkaloids causes inhibition of microtubule assembly through tubulin interaction and prevents cell division. Drugs targeting microtubules causes termination of cell cycle and finally result in cell death through apoptosis. Vinblastine and vincristine has been approved by the FDA for the treatment of various cancers. Taxanes are used for treatment of breast, ovary, lung and other metastatic cancers. Taxanes (paclitaxel, docetaxel) are derived from *Taxus brevifolia*. Mechanism of action this herbal complex was discovered in 1979 [27]. In 2005, Paclitaxel was approved by FDA for clinical usage to treat metastatic breast cancer. Epipodophyllotoxin is an isomer of podophyllotoxin extracted from the root of *Podophyllum peltatum*. Etoposide and teniposide are two active and semi-synthetic compounds of epipodophyllotoxin that show anti-tumor effect through targeting topoisomerae II [106].

Flavopiridol is completely synthetic, but its base is a natural product that is derived from Indian plant *dysoxylum binectariferum*. Flavopiridol causes potent and specific *in vitro* inhibition of all cyclin-dependent kinases (Cdk) and prevents cell cycle progression. Flavopiridol have capacity to induce apoptosis, inhibit angiogenesis, promote differentiation, and modulate transcriptional processes [107]. The crystal structure of Cdk2-flavopiridol complexes reveals that the drug interacts with the ATP-binding site and competes with ATP for its binding with Cdk2 [108]. There are many examples related to successful application of natural products and its derivative as drug for cancer therapy.

Combretastatin A4 is derived from *Combretum caffrum* and inhibit microtubule polymerization, induces G2-M phase cell cycle arrest, and also induces caspase-3 activation [109]. Combretastatin A4 causes cell death mediated through caspase- and non-caspase-dependent pathways. Combretastatin A4 poor water solubility, which makes it unsuitable for intravenous application in bladder cancer. Combretastatin A4 phosphate, a water-soluble form has been used for clinical trials [110]. Combretastatin A4 and their derivatives have shown cytotoxic potency and anti-proliferative activity in non-small cell lung cancer, ovarian cancer, and breast cancer cells. Combretastatin A4 inhibits thyroid cancer cell proliferation, migration and invasion [111]. Combretastatin A4 can be used as a lead compound for development of more potential drug specific to the target. Combretastatin A4 possess conformational instability due to presence of ethylene linker, which joins two aromatic rings. Simplified biphenyl derivatives of combretastatin A4 were designed which lacks an ethylene linker [112]. These biphenyl derivatives of combretastatin A4 has shown similar mechanism of action for cancer therapy.

Roscovitine is another synthetic agent based on a natural product derived from *Raphanus sativus L* [113]. Roscovitine inhibits RNA polymerase II-mediated transcription, which down regulates the expression of anti-apoptotic proteins [114]. Roscovitine was shown to be a competitive inhibitor of p34cdc2/cyclin B kinase. Purine region of the roscovitine molecule binds to the adenine binding pocket of CDK2. Now, R-roscovitine (seliciclib or CYC202) is an orally available cyclin-dependent kinase (CDK) inhibitors that inhibit multiple enzyme targets including CDK2, CDK7 and CDK9 [115].

The new compounds from organisms of diverse biological origin, with potential use as cancer chemotherapeutic agents should be investigated. A large number of compounds with similar biological activity can be used to determine a structure-activity relationships and of its potential mechanism of action. There is need of collaboration for isolation, bioassay, synthesis, optimization, and development of promising new molecules for preclinical trials [116]. Natural products with novel biochemical mechanisms of action may be of much interest for development of cancer drugs. The more potential analogues or derivative of natural compounds should be designed, and their efficacy against a cancer should be evaluated *in vitro* and *in vivo* model. The findings from the *vitro* and *in vivo* testing must be taken into consideration during structural optimization of inhibitors.

CONCLUSION

Most of the drugs derived from the natural sources especially from plants. Therefore, different part of the plants can be investigated in search of lead compounds for cancer prevention and therapy. The molecular targets for different types of natural products should be scientifically validated and must be catalogued with well documented structural, functional and clinical data. Computer aided drug designing approaches can help in developing the more potential, specific and efficacious drugs for a specific type of cancer. Lead optimization of natural compounds or existing inhibitors can improve the ADMET properties. Application of pharmacogenomics principles in cancer therapy can help in reducing the side effects of drugs. Combining various approaches of cancer treatment at different stages along with chemotherapy can produce some impressive results.

CONSENT FOR PUBLICATION

Not applicable.

CONFLICT OF INTEREST

The author(s) confirm that this chapter content has no conflict of interest.

AACKNOWLEDGEMENTS

Declared none

REFERENCES

[1]   Thun, M.J.; DeLancey, J.O.; Center, M.M.; Jemal, A.; Ward, E.M. The global burden of cancer: priorities for prevention. *Carcinogenesis,* **2010**, *31*(1), 100-110.
       [http://dx.doi.org/10.1093/carcin/bgp263] [PMID: 19934210]

[2]   Greenwell, M.; Rahman, P.K.S.M. Medicinal plants: Their use in anticancer treatment. *Int. J. Pharm. Sci. Res.,* **2015**, *6*(10), 4103-4112.
       [PMID: 26594645]

[3]   Torgovnick, A.; Schumacher, B. DNA repair mechanisms in cancer development and therapy. *Front. Genet.,* **2015**, *6*, 157.
       [http://dx.doi.org/10.3389/fgene.2015.00157] [PMID: 25954303]

[4]   Nwodo, J.N.; Ibezim, A.; Simoben, C.V.; Ntie-Kang, F. Exploring cancer therapeutics with natural products from african medicinal plants, part ii: alkaloids, terpenoids and flavonoids. *Anticancer.*

Agents Med. Chem., **2016**, *16*(1), 108-127.
[http://dx.doi.org/10.2174/1871520615666150520143827] [PMID: 25991425]

[5]   Bertoli, C.; Skotheim, J.M.; de Bruin, R.A.M. Control of cell cycle transcription during G1 and S phases. *Nat. Rev. Mol. Cell Biol.,* **2013**, *14*(8), 518-528.
[http://dx.doi.org/10.1038/nrm3629] [PMID: 23877564]

[6]   Davis, R.J. Transcriptional regulation by MAP kinases. *Mol. Reprod. Dev.,* **1995**, *42*(4), 459-467.
[http://dx.doi.org/10.1002/mrd.1080420414] [PMID: 8607977]

[7]   Hachem, A.; Gartenhaus, R.B. Oncogenes as molecular targets in lymphoma. *Blood,* **2005**, *106*(6), 1911-1923.
[http://dx.doi.org/10.1182/blood-2004-12-4621] [PMID: 15802530]

[8]   Valvezan, A.J.; Zhang, F.; Diehl, J.A.; Klein, P.S. Adenomatous polyposis coli (APC) regulates multiple signaling pathways by enhancing glycogen synthase kinase-3 (GSK-3) activity. *J. Biol. Chem.,* **2012**, *287*(6), 3823-3832.
[http://dx.doi.org/10.1074/jbc.M111.323337] [PMID: 22184111]

[9]   Godet, I.; Gilkes, D.M. BRCA1 and BRCA2 mutations and treatment strategies for breast cancer. *Integr. Cancer Sci. Ther.,* **2017**, *4*(1)
[http://dx.doi.org/10.15761/ICST.1000228] [PMID: 28706734]

[10]   Jafri, M.A.; Ansari, S.A.; Alqahtani, M.H.; Shay, J.W. Roles of telomeres and telomerase in cancer, and advances in telomerase-targeted therapies. *Genome Med.,* **2016**, *8*(1), 69.
[http://dx.doi.org/10.1186/s13073-016-0324-x] [PMID: 27323951]

[11]   Singh, S.; Singh, D.B.; Singh, S.; Shukla, R.; Ramteke, P.W.; Misra, K. Exploring medicinal plant legacy for drug discovery in post-genomic era. *Proc. Natl. Acad. Sci., India, Sect. B Biol. Sci.,* **2018**.
[http://dx.doi.org/10.1007/s40011-018-1013-x]

[12]   Bishayee, A.; Sethi, G. Bioactive natural products in cancer prevention and therapy: Progress and promise. *Semin. Cancer Biol.,* **2016**, *40-41*, 1-3.
[http://dx.doi.org/10.1016/j.semcancer.2016.08.006] [PMID: 27565447]

[13]   Patrick, G.L. *An introduction to medicinal chemistry*; Oxford University Press: Oxford, **2013**.

[14]   Bent, S.; Ko, R. Commonly used herbal medicines in the United States: a review. *Am. J. Med.,* **2004**, *116*(7), 478-485.
[http://dx.doi.org/10.1016/j.amjmed.2003.10.036] [PMID: 15047038]

[15]   Yang, H.Y.; Chen, P.C.; Wang, J.D. Chinese herbs containing aristolochic acid associated with renal failure and urothelial carcinoma: a review from epidemiologic observations to causal inference. *BioMed Res. Int.,* **2014**, *2014*569325
[http://dx.doi.org/10.1155/2014/569325] [PMID: 25431765]

[16]   Chen, T.; Guo, L.; Zhang, L.; Shi, L.; Fang, H.; Sun, Y.; Fuscoe, J.C.; Mei, N. Gene expression profiles distinguish the carcinogenic effects of aristolochic acid in target (kidney) and non-target (liver) tissues in rats. *BMC Bioinformatics,* **2006**, *7*(2) Suppl. 2, S20.
[http://dx.doi.org/10.1186/1471-2105-7-S2-S20] [PMID: 17118142]

[17]   Singh, D.B. Pharmacogenomics: clinical perspective, strategies, and challenges. In: *Translational bioinformatics and its application*; Springer: Dordrecht, **2017**; pp. 299-333.
[http://dx.doi.org/10.1007/978-94-024-1045-7_13]

[18]   Shanmugam, M.K.; Rane, G.; Kanchi, M.M.; Arfuso, F.; Chinnathambi, A.; Zayed, M.E.; Alharbi, S.A.; Tan, B.K.; Kumar, A.P.; Sethi, G. The multifaceted role of curcumin in cancer prevention and treatment. *Molecules,* **2015**, *20*(2), 2728-2769.
[http://dx.doi.org/10.3390/molecules20022728] [PMID: 25665066]

[19]   He, D.; Xu, Q.; Yan, M.; Zhang, P.; Zhou, X.; Zhang, Z.; Duan, W.; Zhong, L.; Ye, D.; Chen, W. The NF-kappa B inhibitor, celastrol, could enhance the anti-cancer effect of gambogic acid on oral squamous cell carcinoma. *BMC Cancer,* **2009**, *9*, 343.

[http://dx.doi.org/10.1186/1471-2407-9-343] [PMID: 19778460]

[20]   Shanmugam, M.K.; Lee, J.H.; Chai, E.Z.; Kanchi, M.M.; Kar, S.; Arfuso, F.; Dharmarajan, A.; Kumar, A.P.; Ramar, P.S.; Looi, C.Y.; Mustafa, M.R.; Tergaonkar, V.; Bishayee, A.; Ahn, K.S.; Sethi, G. Cancer prevention and therapy through the modulation of transcription factors by bioactive natural compounds. *Semin. Cancer Biol.,* **2016**, *40-41*, 35-47.
[http://dx.doi.org/10.1016/j.semcancer.2016.03.005] [PMID: 27038646]

[21]   Yarla, N.S.; Bishayee, A.; Sethi, G.; Reddanna, P.; Kalle, A.M.; Dhananjaya, B.L.; Dowluru, K.S.; Chintala, R.; Duddukuri, G.R. Targeting arachidonic acid pathway by natural products for cancer prevention and therapy. *Semin. Cancer Biol.,* **2016**, *40-41*, 48-81.
[http://dx.doi.org/10.1016/j.semcancer.2016.02.001] [PMID: 26853158]

[22]   Ong, W.Y.; Farooqui, T.; Kokotos, G.; Farooqui, A.A. Synthetic and natural inhibitors of phospholipases A2: their importance for understanding and treatment of neurological disorders. *ACS Chem. Neurosci.,* **2015**, *6*(6), 814-831.
[http://dx.doi.org/10.1021/acschemneuro.5b00073] [PMID: 25891385]

[23]   Barbosa, N.R.; Pittella, F.; Gattaz, W.F. Centella asiatica water extract inhibits iPLA2 and cPLA2 activities in rat cerebellum. *Phytomedicine,* **2008**, *15*(10), 896-900.
[http://dx.doi.org/10.1016/j.phymed.2008.02.007] [PMID: 18455381]

[24]   Bak, M.J.; Das Gupta, S.; Wahler, J.; Suh, N. Role of dietary bioactive natural products in estrogen receptor-positive breast cancer. *Semin. Cancer Biol.,* **2016**, *40-41*, 170-191.
[http://dx.doi.org/10.1016/j.semcancer.2016.03.001] [PMID: 27016037]

[25]   Jäger, K.; Walter, M. Therapeutic Targeting of Telomerase. *Genes (Basel),* **2016**, *7*(7)E39
[http://dx.doi.org/10.3390/genes7070039] [PMID: 27455328]

[26]   Lee, J.H.; Chung, I.K. Curcumin inhibits nuclear localization of telomerase by dissociating the Hsp90 co-chaperone p23 from hTERT. *Cancer Lett.,* **2010**, *290*(1), 76-86.
[http://dx.doi.org/10.1016/j.canlet.2009.08.026] [PMID: 19751963]

[27]   Safarzadeh, E.; Sandoghchian Shotorbani, S.; Baradaran, B. Herbal medicine as inducers of apoptosis in cancer treatment. *Adv. Pharm. Bull.,* **2014**, *4* Suppl. 1, 421-427.
[PMID: 25364657]

[28]   Lin, S.R.; Fu, Y.S.; Tsai, M.J.; Cheng, H.; Weng, C.F. Natural Compounds from Herbs that can Potentially Execute as Autophagy Inducers for Cancer Therapy. *Int. J. Mol. Sci.,* **2017**, *18*(7)E1412
[http://dx.doi.org/10.3390/ijms18071412] [PMID: 28671583]

[29]   Kadioglu, O.; Cao, J.; Saeed, M.E.; Greten, H.J.; Efferth, T. Targeting epidermal growth factor receptors and downstream signaling pathways in cancer by phytochemicals. *Target. Oncol.,* **2015**, *10*(3), 337-353.
[http://dx.doi.org/10.1007/s11523-014-0339-4] [PMID: 25410594]

[30]   Zhou, J.; Giannakakou, P. Targeting microtubules for cancer chemotherapy. *Curr. Med. Chem. Anticancer Agents,* **2005**, *5*(1), 65-71.
[http://dx.doi.org/10.2174/1568011053352569] [PMID: 15720262]

[31]   Jain, C.K.; Majumder, H.K.; Roychoudhury, S. Natural Compounds as Anticancer Agents Targeting DNA Topoisomerases. *Curr. Genomics,* **2017**, *18*(1), 75-92.
[http://dx.doi.org/10.2174/1389202917666160808125213] [PMID: 28503091]

[32]   Song, X.; Xiong, Y.; Qi, X.; Tang, W.; Dai, J.; Gu, Q.; Li, J. Molecular Targets of Active Anticancer Compounds Derived from Marine Sources. *Mar. Drugs,* **2018**, *16*(5), 175.
[http://dx.doi.org/10.3390/md16050175] [PMID: 29786660]

[33]   Ashida, Y.; Yanagita, R.C.; Takahashi, C.; Kawanami, Y.; Irie, K. Binding mode prediction of aplysiatoxin, a potent agonist of protein kinase C, through molecular simulation and structure-activity study on simplified analogs of the receptor-recognition domain. *Bioorg. Med. Chem.,* **2016**, *24*(18), 4218-4227.

[http://dx.doi.org/10.1016/j.bmc.2016.07.011] [PMID: 27436807]

[34]   Zovko, A.; Viktorsson, K.; Hååg, P.; Kovalerchick, D.; Färnegårdh, K.; Alimonti, A.; Ilan, M.; Carmeli, S.; Lewensohn, R. Marine sponge Cribrochalina vasculum compounds activate intrinsic apoptotic signaling and inhibit growth factor signaling cascades in non-small cell lung carcinoma. *Mol. Cancer Ther.*, **2014**, *13*(12), 2941-2954. [http://dx.doi.org/10.1158/1535-7163.MCT-14-0329] [PMID: 25319389]

[35]   Sauleau, P.; Retailleau, P.; Nogues, S.; Carletti, I.; Marcourt, L.; Raux, R.; Mourabit, A.A.; Debitus, C. Dihydrohymenialdisines, new pyrrole-2-aminoimidazole alkaloids from the marine sponge Cymbastela cantharella. *Tetrahedron Lett.*, **2011**, *52*, 2676-2678. [http://dx.doi.org/10.1016/j.tetlet.2011.03.073]

[36]   Lin, J.; Yan, X-J.; Chen, H-M. Fascaplysin, a selective CDK4 inhibitor, exhibit anti-angiogenic activity *in vitro* and *in vivo. Cancer Chemother. Pharmacol.*, **2007**, *59*(4), 439-445. [http://dx.doi.org/10.1007/s00280-006-0282-x] [PMID: 16816972]

[37]   Bharate, S.B.; Yadav, R.R.; Battula, S.; Vishwakarma, R.A. Meridianins: marine-derived potent kinase inhibitors. *Mini Rev. Med. Chem.*, **2012**, *12*(7), 618-631. [http://dx.doi.org/10.2174/138955712800626728] [PMID: 22512550]

[38]   Bidon-Chanal, A.; Fuertes, A.; Alonso, D.; Pérez, D.I.; Martínez, A.; Luque, F.J.; Medina, M. Evidence for a new binding mode to GSK-3: allosteric regulation by the marine compound palinurin. *Eur. J. Med. Chem.*, **2013**, *60*, 479-489. [http://dx.doi.org/10.1016/j.ejmech.2012.12.014] [PMID: 23354070]

[39]   Guzmán, E.A.; Johnson, J.D.; Linley, P.A.; Gunasekera, S.E.; Wright, A.E. A novel activity from an old compound: Manzamine A reduces the metastatic potential of AsPC-1 pancreatic cancer cells and sensitizes them to TRAIL-induced apoptosis. *Invest. New Drugs*, **2011**, *29*(5), 777-785. [http://dx.doi.org/10.1007/s10637-010-9422-6] [PMID: 20352293]

[40]   Sallam, A.A.; Mohyeldin, M.M.; Foudah, A.I.; Akl, M.R.; Nazzal, S.; Meyer, S.A.; Liu, Y-Y.; El Sayed, K.A. Marine natural products-inspired phenylmethylene hydantoins with potent *in vitro* and *in vivo* antitumor activities *via* suppression of Brk and FAK signaling. *Org. Biomol. Chem.*, **2014**, *12*(28), 5295-5303. [http://dx.doi.org/10.1039/C4OB00553H] [PMID: 24927150]

[41]   Wiese, J.; Imhoff, J.F.; Gulder, T.A.; Labes, A.; Schmaljohann, R. Marine Fungi as Producers of Benzocoumarins, a New Class of Inhibitors of Glycogen-Synthase-Kinase 3β. *Mar. Drugs*, **2016**, *14*(11), 200. [http://dx.doi.org/10.3390/md14110200] [PMID: 27801816]

[42]   Wätjen, W.; Ebada, S.S.; Bergermann, A.; Chovolou, Y.; Totzke, F.; Kubbutat, M.H.G.; Lin, W.; Proksch, P. Cytotoxic effects of the anthraquinone derivatives 1′-deoxyrhodoptilometrin and (S)-(-)-rhodoptilometrin isolated from the marine echinoderm Comanthus sp. *Arch. Toxicol.*, **2017**, *91*(3), 1485-1495. [http://dx.doi.org/10.1007/s00204-016-1787-7] [PMID: 27473261]

[43]   Wu, N.; Luo, J.; Jiang, B.; Wang, L.; Wang, S.; Wang, C.; Fu, C.; Li, J.; Shi, D. Marine bromophenol bis (2,3-dibromo-4,5-dihydroxy-phenyl)-methane inhibits the proliferation, migration, and invasion of hepatocellular carcinoma cells *via* modulating β1-integrin/FAK signaling. *Mar. Drugs*, **2015**, *13*(2), 1010-1025. [http://dx.doi.org/10.3390/md13021010] [PMID: 25689564]

[44]   Goey, A.K.L.; Chau, C.H.; Sissung, T.M.; Cook, K.M.; Venzon, D.J.; Castro, A.; Ransom, T.R.; Henrich, C.J.; McKee, T.C.; McMahon, J.B.; Grkovic, T.; Cadelis, M.M.; Copp, B.R.; Gustafson, K.R.; Figg, W.D. Screening and Biological Effects of Marine Pyrroloiminoquinone Alkaloids: Potential Inhibitors of the HIF-1α/p300 Interaction. *J. Nat. Prod.*, **2016**, *79*(5), 1267-1275. [http://dx.doi.org/10.1021/acs.jnatprod.5b00846] [PMID: 27140429]

[45]   Baud, M.G.J.; Leiser, T.; Haus, P.; Samlal, S.; Wong, A.C.; Wood, R.J.; Petrucci, V.; Gunaratnam,

M.; Hughes, S.M.; Buluwela, L.; Turlais, F.; Neidle, S.; Meyer-Almes, F.J.; White, A.J.; Fuchter, M.J. Defining the mechanism of action and enzymatic selectivity of psammaplin A against its epigenetic targets. *J. Med. Chem.,* **2012**, *55*(4), 1731-1750.
[http://dx.doi.org/10.1021/jm2016182] [PMID: 22280363]

[46]   Liu, Y.; Salvador, L.A.; Byeon, S.; Ying, Y.; Kwan, J.C.; Law, B.K.; Hong, J.; Luesch, H. Anticolon cancer activity of largazole, a marine-derived tunable histone deacetylase inhibitor. *J. Pharmacol. Exp. Ther.,* **2010**, *335*(2), 351-361.
[http://dx.doi.org/10.1124/jpet.110.172387] [PMID: 20739454]

[47]   Song, X.; Zhao, Z.; Qi, X.; Tang, S.; Wang, Q.; Zhu, T.; Gu, Q.; Liu, M.; Li, J. Identification of epipolythiodioxopiperazines HDN-1 and chaetocin as novel inhibitor of heat shock protein 90. *Oncotarget,* **2015**, *6*(7), 5263-5274.
[http://dx.doi.org/10.18632/oncotarget.3029] [PMID: 25742791]

[48]   Lai, K.H.; Liu, Y.C.; Su, J.H.; El-Shazly, M.; Wu, C.F.; Du, Y.C.; Hsu, Y.M.; Yang, J.C.; Weng, M.K.; Chou, C.H.; Chen, G.Y.; Chen, Y.C.; Lu, M.C. Antileukemic Scalarane Sesterterpenoids and Meroditerpenoid from Carteriospongia (Phyllospongia) sp., Induce Apoptosis *via* Dual Inhibitory Effects on Topoisomerase II and Hsp90. *Sci. Rep.,* **2016**, *6*, 36170.
[http://dx.doi.org/10.1038/srep36170] [PMID: 27796344]

[49]   Afifi, A.H.; Kagiyama, I.; El-Desoky, A.H.; Kato, H.; Mangindaan, R.E.P.; de Voogd, N.J.; Ammar, N.M.; Hifnawy, M.S.; Tsukamoto, S.; Sulawesins, A.C. Sulawesins A-C, Furanosesterterpene Tetronic Acids That Inhibit USP7, from a Psammocinia sp. Marine Sponge. *J. Nat. Prod.,* **2017**, *80*(7), 2045-2050.
[http://dx.doi.org/10.1021/acs.jnatprod.7b00184] [PMID: 28621941]

[50]   Abraham, I.; Jain, S.; Wu, C.P.; Khanfar, M.A.; Kuang, Y.; Dai, C.L.; Shi, Z.; Chen, X.; Fu, L.; Ambudkar, S.V.; El Sayed, K.; Chen, Z.S. Marine sponge-derived sipholane triterpenoids reverse P-glycoprotein (ABCB1)-mediated multidrug resistance in cancer cells. *Biochem. Pharmacol.,* **2010**, *80*(10), 1497-1506.
[http://dx.doi.org/10.1016/j.bcp.2010.08.001] [PMID: 20696137]

[51]   Hodnik, Ž.; Peterlin Mašič, L.; Tomašić, T.; Smodiš, D.; D'Amore, C.; Fiorucci, S.; Kikelj, D. Bazedoxifene-scaffold-based mimetics of solomonsterols A and B as novel pregnane X receptor antagonists. *J. Med. Chem.,* **2014**, *57*(11), 4819-4833.
[http://dx.doi.org/10.1021/jm500351m] [PMID: 24828006]

[52]   Lin, S.R.; Chang, C.H.; Hsu, C.F.; Tsai, M.J.; Cheng, H.; Leong, M.K.; Sung, P.J.; Chen, J.C.; Weng, C.F. Natural compounds as potential adjuvants to cancer therapy: Preclinical evidence. *Br. J. Pharmacol.,* **2019**.
[http://dx.doi.org/10.1111/bph.14816] [PMID: 31368509]

[53]   Santos, F.M.; Latorre, A.O.; Hueza, I.M.; Sanches, D.S.; Lippi, L.L.; Gardner, D.R.; Spinosa, H.S. Increased antitumor efficacy by the combined administration of swainsonine and cisplatin *in vivo*. *Phytomedicine,* **2011**, *18*(12), 1096-1101.
[http://dx.doi.org/10.1016/j.phymed.2011.06.005] [PMID: 21763115]

[54]   Qi, F.; Li, A.; Inagaki, Y.; Gao, J.; Li, J.; Kokudo, N.; Li, X.K.; Tang, W. Chinese herbal medicines as adjuvant treatment during chemo- or radio-therapy for cancer. *Biosci. Trends,* **2010**, *4*(6), 297-307.
[PMID: 21248427]

[55]   Wang, C.Z.; Calway, T.; Yuan, C.S. Herbal medicines as adjuvants for cancer therapeutics. *Am. J. Chin. Med.,* **2012**, *40*(4), 657-669.
[http://dx.doi.org/10.1142/S0192415X12500498] [PMID: 22809022]

[56]   Ramawat, K.G.; Goyal, S. Natural products in cancer chemoprevention and chemotherapy.*Herbal drugs: ethnomedicine to modern medicine*; Springer: Berlin, Heidelberg, **2009**, pp. 153-171.
[http://dx.doi.org/10.1007/978-3-540-79116-4_10]

[57]   Aung, T.N.; Qu, Z.; Kortschak, R.D.; Adelson, D.L. Understanding the effectiveness of natural

compound mixtures in cancer through their molecular mode of action. *Int. J. Mol. Sci.,* **2017**, *18*(3), 656.
[http://dx.doi.org/10.3390/ijms18030656] [PMID: 28304343]

[58]   Hecht, S.S. Chemoprevention of cancer by isothiocyanates, modifiers of carcinogen metabolism. *J. Nutr.,* **1999**, *129*(3), 768S-774S.
[http://dx.doi.org/10.1093/jn/129.3.768S] [PMID: 10082787]

[59]   Anzai, N.; Taniyama, T.; Nakandakari, N.; Sugiyama, C.; Negishi, T.; Hayatsu, H.; Negishi, K. Inhibition of DNA adduct formation and mutagenic action of 3-amino-1-methyl-5h-pyrid-[4,3-b]indole by chlorophyllin-chitosan in rpsL transgenic mice. *Jpn. J. Cancer Res.,* **2001**, *92*(8), 848-853.
[http://dx.doi.org/10.1111/j.1349-7006.2001.tb01171.x] [PMID: 11509116]

[60]   Surh, Y.J. Cancer chemoprevention with dietary phytochemicals. *Nat. Rev. Cancer,* **2003**, *3*(10), 768-780.
[http://dx.doi.org/10.1038/nrc1189] [PMID: 14570043]

[61]   Weston, A.; Poirier, M.C. *Carcinogen–DNA Adduct Formation and DNA Repair*; , **2005**.
[http://dx.doi.org/10.1016/B0-12-369400-0/00191-5]

[62]   Zhang, Y.; Munday, R. Dithiolethiones for cancer chemoprevention: where do we stand? *Mol. Cancer Ther.,* **2008**, *7*(11), 3470-3479.
[http://dx.doi.org/10.1158/1535-7163.MCT-08-0625] [PMID: 19001432]

[63]   Jubert, C.; Mata, J.; Bench, G.; Dashwood, R.; Pereira, C.; Tracewell, W.; Turteltaub, K.; Williams, D.; Bailey, G. Effects of chlorophyll and chlorophyllin on low-dose aflatoxin B(1) pharmacokinetics in human volunteers. *Cancer Prev. Res. (Phila.),* **2009**, *2*(12), 1015-1022.
[http://dx.doi.org/10.1158/1940-6207.CAPR-09-0099] [PMID: 19952359]

[64]   Raffoul, J.J.; Kucuk, O.; Sarkar, F.H.; Hillman, G.G. Dietary agents in cancer chemoprevention and treatment. *J. Oncol.,* **2012**, *2012*749310
[http://dx.doi.org/10.1155/2012/749310] [PMID: 23316231]

[65]   Patel, R.; Garg, R.; Erande, S.; B Maru, G. Chemopreventive herbal anti-oxidants: current status and future perspectives. *J. Clin. Biochem. Nutr.,* **2007**, *40*(2), 82-91.
[http://dx.doi.org/10.3164/jcbn.40.82] [PMID: 18188409]

[66]   Singh, S.; Sharma, B.; Kanwar, S.S.; Kumar, A. Lead phytochemicals for anticancer drug development. *Front. Plant Sci.,* **2016**, *7*, 1667.
[http://dx.doi.org/10.3389/fpls.2016.01667] [PMID: 27877185]

[67]   Hussain, S.S.; Kumar, A.P.; Ghosh, R. Food-based natural products for cancer management: Is the whole greater than the sum of the parts?*Seminars in cancer biology*; Academic Press, **2016**, 40, pp. 233-246.

[68]   Kesharwani, R.K.; Misra, K.; Singh, D.B. Perspectives and challenges of tropical medicinal herbs and modern drug discovery in the current scenario. *Asian. Pac. J. Trop. Med.,* **2019**, *12*, 1-7.

[69]   Langner, E.; Rzeski, W. Dietary derived compounds in cancer chemoprevention. *Contemp. Oncol. (Pozn.),* **2012**, *16*(5), 394-400.
[http://dx.doi.org/10.5114/wo.2012.31767] [PMID: 23788916]

[70]   Pan, M.H.; Ho, C.T. Chemopreventive effects of natural dietary compounds on cancer development. *Chem. Soc. Rev.,* **2008**, *37*(11), 2558-2574.
[http://dx.doi.org/10.1039/b801558a] [PMID: 18949126]

[71]   Demain, A.L.; Vaishnav, P. Natural products for cancer chemotherapy. *Microb. Biotechnol.,* **2011**, *4*(6), 687-699.
[http://dx.doi.org/10.1111/j.1751-7915.2010.00221.x] [PMID: 21375717]

[72]   Sarkar, F.H.; Li, Y.; Wang, Z.; Kong, D. Cellular signaling perturbation by natural products. *Cell. Signal.,* **2009**, *21*(11), 1541-1547.

[http://dx.doi.org/10.1016/j.cellsig.2009.03.009] [PMID: 19298854]

[73] Balasubramanian, S.; Hurley, L.H.; Neidle, S. Targeting G-quadruplexes in gene promoters: a novel anticancer strategy? *Nat. Rev. Drug Discov.,* **2011**, *10*(4), 261-275.
[http://dx.doi.org/10.1038/nrd3428] [PMID: 21455236]

[74] Di Somma, S.; Amato, J.; Iaccarino, N.; Pagano, B.; Randazzo, A.; Portella, G.; Malfitano, A.M. G-Quadruplex Binders Induce Immunogenic Cell Death Markers in Aggressive Breast Cancer Cells. *Cancers (Basel),* **2019**, *11*(11), 1797.
[http://dx.doi.org/10.3390/cancers11111797] [PMID: 31731707]

[75] Mann, J. Natural products in cancer chemotherapy: past, present and future. *Nat. Rev. Cancer,* **2002**, *2*(2), 143-148.
[http://dx.doi.org/10.1038/nrc723] [PMID: 12635177]

[76] Newman, D.J.; Cragg, G.M. Natural products as sources of new drugs over the 30 years from 1981 to 2010. *J. Nat. Prod.,* **2012**, *75*(3), 311-335.
[http://dx.doi.org/10.1021/np200906s] [PMID: 22316239]

[77] Cragg, G.M.; Kingston, D.G.I.; Newman, D.J. *Anticancer Agents from Natural Products,* 2nd ed; CRC: Boca Raton, **2012**.

[78] Cragg, G.M.; Pezzuto, J.M. Natural products as a vital source for the discovery of cancer chemotherapeutic and chemopreventive agents. *Med. Princ. Pract.,* **2016**, *25* Suppl. 2, 41-59.
[http://dx.doi.org/10.1159/000443404] [PMID: 26679767]

[79] Newman, D.J.; Cragg, G.M. Marine-sourced anti-cancer and cancer pain control agents in clinical and late preclinical development. *Mar. Drugs,* **2014**, *12*(1), 255-278.
[http://dx.doi.org/10.3390/md12010255] [PMID: 24424355]

[80] Zhang, Q.Y.; Wang, F.X.; Jia, K.K.; Kong, L.D. Natural product interventions for chemotherapy and radiotherapy-induced side effects. *Front. Pharmacol.,* **2018**, *9*, 1253.
[http://dx.doi.org/10.3389/fphar.2018.01253] [PMID: 30459615]

[81] Song, C.; Li, F.; Wang, S.; Wang, J.; Wei, W.; Ma, G. Recent Advances in Particulate Adjuvants for Cancer Vaccination. *Adv. Ther.,* **2019**.
[http://dx.doi.org/10.1002/adtp.201900115]

[82] Leow, P.C.; Tian, Q.; Ong, Z.Y.; Yang, Z.; Ee, P.L. Antitumor activity of natural compounds, curcumin and PKF118-310, as Wnt/β-catenin antagonists against human osteosarcoma cells. *Invest. New Drugs,* **2010**, *28*(6), 766-782.
[http://dx.doi.org/10.1007/s10637-009-9311-z] [PMID: 19730790]

[83] Sagar, M.; Pathak, R.K.; Pandey, R.K.; Singh, D.B.; Pandey, N.; Gupta, M.K. Binding affinity analysis and ADMET prediction of epigallocatechine gallate (EGCG) derivatives for AP-1 protein: a drug International Journal of Peptide Research and Therapeutics 1 3 target for liver cancer. *Netw. Model. Anal. Health Inform. Bioinform.,* **2014**, *3*, 66.
[http://dx.doi.org/10.1007/s13721-014-0066-x]

[84] Kudo, I.; Murakami, M. Phospholipase A2 enzymes. *Prostaglandins Other Lipid Mediat.,* **2002**, *68-69*, 3-58.
[http://dx.doi.org/10.1016/S0090-6980(02)00020-5] [PMID: 12432908]

[85] Yue, W.; Wang, J-P.; Li, Y.; Fan, P.; Liu, G.; Zhang, N.; Conaway, M.; Wang, H.; Korach, K.S.; Bocchinfuso, W.; Santen, R. Effects of estrogen on breast cancer development: Role of estrogen receptor independent mechanisms. *Int. J. Cancer,* **2010**, *127*(8), 1748-1757.
[http://dx.doi.org/10.1002/ijc.25207] [PMID: 20104523]

[86] Burchett, K.M.; Yan, Y.; Ouellette, M.M. Telomerase inhibitor Imetelstat (GRN163L) limits the lifespan of human pancreatic cancer cells. *PLoS One,* **2014**, *9*(1)e85155
[http://dx.doi.org/10.1371/journal.pone.0085155] [PMID: 24409321]

[87] Cheng, S.; Gao, N.; Zhang, Z.; Chen, G.; Budhraja, A.; Ke, Z.; Son, Y.O.; Wang, X.; Luo, J.; Shi, X.

Quercetin induces tumor-selective apoptosis through downregulation of Mcl-1 and activation of Bax. *Clin. Cancer Res.,* **2010**, *16*(23), 5679-5691.
[http://dx.doi.org/10.1158/1078-0432.CCR-10-1565] [PMID: 21138867]

[88]   Tiash, S.; Chowdhury, E. Growth factor receptors: Promising drug targets in cancer. *J. Cancer Metastasis Treat.,* **2015**, *1*, 190-200.
[http://dx.doi.org/10.4103/2394-4722.163151]

[89]   Wee, P.; Wang, Z. Epidermal Growth Factor Receptor Cell Proliferation Signaling Pathways. *Cancers (Basel),* **2017**, *9*(5)E52
[PMID: 28513565]

[90]   Perea, S.; Hidalgo, M. Predictors of sensitivity and resistance to epidermal growth factor receptor inhibitors. *Clin. Lung Cancer,* **2004**, *6*(1) Suppl. 1, S30-S34.
[http://dx.doi.org/10.3816/CLC.2004.s.012] [PMID: 15638955]

[91]   Patel, B.B.; Gupta, D.; Elliott, A.A.; Sengupta, V.; Yu, Y.; Majumdar, A.P. Curcumin targets FOLFOX-surviving colon cancer cells *via* inhibition of EGFRs and IGF-1R. *Anticancer Res.,* **2010**, *30*(2), 319-325.
[PMID: 20332435]

[92]   Lee, M.F.; Pan, M.H.; Chiou, Y.S.; Cheng, A.C.; Huang, H. Resveratrol modulates MED28 (Magicin/EG-1) expression and inhibits epidermal growth factor (EGF)-induced migration in MDA-MB-231 human breast cancer cells. *J. Agric. Food Chem.,* **2011**, *59*(21), 11853-11861.
[http://dx.doi.org/10.1021/jf202426k] [PMID: 21942447]

[93]   Nigg, E.A.; Raff, J.W. Centrioles, centrosomes, and cilia in health and disease. *Cell,* **2009**, *139*(4), 663-678.
[http://dx.doi.org/10.1016/j.cell.2009.10.036] [PMID: 19914163]

[94]   Landen, J.W.; Lang, R.; McMahon, S.J.; Rusan, N.M.; Yvon, A.M.; Adams, A.W.; Sorcinelli, M.D.; Campbell, R.; Bonaccorsi, P.; Ansel, J.C.; Archer, D.R.; Wadsworth, P.; Armstrong, C.A.; Joshi, H.C. Noscapine alters microtubule dynamics in living cells and inhibits the progression of melanoma. *Cancer Res.,* **2002**, *62*(14), 4109-4114.
[PMID: 12124349]

[95]   Kathiravan, M.K.; Kale, A.N.; Nilewar, S. Discovery and Development of Topoisomerase Inhibitors as Anticancer Agents. *Mini Rev. Med. Chem.,* **2016**, *16*(15), 1219-1229.
[http://dx.doi.org/10.2174/1389557516666160822110819] [PMID: 27549098]

[96]   Wu, N.; Wu, X.W.; Agama, K.; Pommier, Y.; Du, J.; Li, D.; Gu, L.Q.; Huang, Z.S.; An, L.K. A novel DNA topoisomerase I inhibitor with different mechanism from camptothecin induces G2/M phase cell cycle arrest to K562 cells. *Biochemistry,* **2010**, *49*(47), 10131-10136.
[http://dx.doi.org/10.1021/bi1009419] [PMID: 21033700]

[97]   Baldwin, E.L.; Osheroff, N. Etoposide, topoisomerase II and cancer. *Curr. Med. Chem. Anticancer Agents,* **2005**, *5*(4), 363-372.
[http://dx.doi.org/10.2174/1568011054222364] [PMID: 16101488]

[98]   Reed, D.; Raina, K.; Agarwal, R. Nutraceuticals in prostate cancer therapeutic strategies and their neo-adjuvant use in diverse populations. *NPJ Precis Oncol,* **2018**, *2*, 15.
[http://dx.doi.org/10.1038/s41698-018-0058-x] [PMID: 30062144]

[99]   Salami, A.; Seydi, E.; Pourahmad, J. Use of nutraceuticals for prevention and treatment of cancer. *Iran. J. Pharm. Res.,* **2013**, *12*(3), 219-220.
[PMID: 24250626]

[100] Saldanha, S.N.; Tollefsbol, T.O. The role of nutraceuticals in chemoprevention and chemotherapy and their clinical outcomes. *J. Oncol.,* **2012**, *2012*192464
[http://dx.doi.org/10.1155/2012/192464] [PMID: 22187555]

[101] Rabi, T.; Bishayee, A. d -Limonene sensitizes docetaxel-induced cytotoxicity in human prostate cancer

cells: Generation of reactive oxygen species and induction of apoptosis. *J. Carcinog.*, **2009**, *8*, 9.
[http://dx.doi.org/10.4103/1477-3163.51368] [PMID: 19465777]

[102]   Mukhtar, Y.M.; Adu-Frimpong, M.; Xu, X.; Yu, J. Biochemical significance of limonene and its metabolites: future prospects for designing and developing highly potent anticancer drugs. *Biosci. Rep.*, **2018**, *38*(6)BSR20181253
[http://dx.doi.org/10.1042/BSR20181253] [PMID: 30287506]

[103]   Chen, T.C.; Fonseca, C.O.; Schönthal, A.H. Preclinical development and clinical use of perillyl alcohol for chemoprevention and cancer therapy. *Am. J. Cancer Res.*, **2015**, *5*(5), 1580-1593.
[PMID: 26175929]

[104]   Zhang, C.; Zhai, S.; Li, X.; Zhang, Q.; Wu, L.; Liu, Y.; Jiang, C.; Zhou, H.; Li, F.; Zhang, S.; Su, G.; Zhang, B.; Yan, B. Synergistic action by multi-targeting compounds produces a potent compound combination for human NSCLC both *in vitro* and *in vivo*. *Cell Death Dis.*, **2014**, *5*e1138
[http://dx.doi.org/10.1038/cddis.2014.76] [PMID: 24651441]

[105]   Moudi, M.; Go, R.; Yien, C.Y.; Nazre, M. Vinca alkaloids. *Int. J. Prev. Med.*, **2013**, *4*(11), 1231-1235.
[PMID: 24404355]

[106]   Ketron, A.C.; Osheroff, N. Phytochemicals as Anticancer and Chemopreventive Topoisomerase II Poisons. *Phytochem. Rev.*, **2014**, *13*(1), 19-35.
[http://dx.doi.org/10.1007/s11101-013-9291-7] [PMID: 24678287]

[107]   Senderowicz, A.M. Flavopiridol: the first cyclin-dependent kinase inhibitor in human clinical trials. *Invest. New Drugs,* **1999**, *17*(3), 313-320.
[http://dx.doi.org/10.1023/A:1006353008903] [PMID: 10665481]

[108]   Chao, S.H.; Price, D.H. Flavopiridol inactivates P-TEFb and blocks most RNA polymerase II transcription *in vivo*. *J. Biol. Chem.*, **2001**, *276*(34), 31793-31799.
[http://dx.doi.org/10.1074/jbc.M102306200] [PMID: 11431468]

[109]   Shen, C.H.; Shee, J.J.; Wu, J.Y.; Lin, Y.W.; Wu, J.D.; Liu, Y.W. Combretastatin A-4 inhibits cell growth and metastasis in bladder cancer cells and retards tumour growth in a murine orthotopic bladder tumour model. *Br. J. Pharmacol.*, **2010**, *160*(8), 2008-2027.
[http://dx.doi.org/10.1111/j.1476-5381.2010.00861.x] [PMID: 20649598]

[110]   Stevenson, J.P.; Rosen, M.; Sun, W.; Gallagher, M.; Haller, D.G.; Vaughn, D.; Giantonio, B.; Zimmer, R.; Petros, W.P.; Stratford, M.; Chaplin, D.; Young, S.L.; Schnall, M.; O'Dwyer, P.J. Phase I trial of the antivascular agent combretastatin A4 phosphate on a 5-day schedule to patients with cancer: magnetic resonance imaging evidence for altered tumor blood flow. *J. Clin. Oncol.*, **2003**, *21*(23), 4428-4438.
[http://dx.doi.org/10.1200/JCO.2003.12.986] [PMID: 14645433]

[111]   Liang, W.; Lai, Y.; Zhu, M.; Huang, S.; Feng, W.; Gu, X. Combretastatin A4 Regulates Proliferation, Migration, Invasion, and Apoptosis of Thyroid Cancer Cells *via* PI3K/Akt Signaling Pathway. *Med. Sci. Monit.*, **2016**, *22*, 4911-4917.
[http://dx.doi.org/10.12659/MSM.898545] [PMID: 27966519]

[112]   Tarade, D.; Ma, D.; Pignanelli, C.; Mansour, F.; Simard, D.; van den Berg, S.; Gauld, J.; McNulty, J.; Pandey, S. Structurally simplified biphenyl combretastatin A4 derivatives retain *in vitro* anti-cancer activity dependent on mitotic arrest. *PLoS One*, **2017**, *12*(3)e0171806
[http://dx.doi.org/10.1371/journal.pone.0171806] [PMID: 28253265]

[113]   Cragg, G.M.; Newman, D.J. Plants as a source of anti-cancer agents. *J. Ethnopharmacol.*, **2005**, *100*(1-2), 72-79.
[http://dx.doi.org/10.1016/j.jep.2005.05.011] [PMID: 16009521]

[114]   Li, L.; Wang, H.; Kim, Js.; Pihan, G.; Boussiotis, V. The cyclin dependent kinase inhibitor (R)-roscovitine prevents alloreactive T cell clonal expansion and protects against acute GvHD. *Cell Cycle,* **2009**, *8*(11), 1794-1802.
[http://dx.doi.org/10.4161/cc.8.11.8738] [PMID: 19448431]

[115]   MacCallum, D.E.; Melville, J.; Frame, S.; Watt, K.; Anderson, S.; Gianella-Borradori, A.; Lane, D.P.; Green, S.R. Seliciclib (CYC202, R-Roscovitine) induces cell death in multiple myeloma cells by inhibition of RNA polymerase II-dependent transcription and down-regulation of Mcl-1. *Cancer Res.,* **2005**, *65*(12), 5399-5407.
[http://dx.doi.org/10.1158/0008-5472.CAN-05-0233] [PMID: 15958589]

[116]   Kinghorn, A.D.; Chin, Y.W.; Swanson, S.M. Discovery of natural product anticancer agents from biodiverse organisms. *Curr. Opin. Drug Discov. Devel.,* **2009**, *12*(2), 189-196.
[PMID: 19333864]

SUBJECT INDEX

A

Absorption 30, 156
  glucose 30
ABTS 19, 27
  assay 19
  radicals scavenging assays 27
ACE 10, 15, 19, 25, 28, 29
  activity 15, 19, 25
   inhibited 29
   substrates 28
ACEi 9, 10, 13, 14, 15, 18, 19, 24, 25, 29
  activity 9, 10, 13, 14, 15, 18, 19, 24, 25, 29
  fraction 25
  peptides 29
ACE inhibitor(s) 9, 28, 29
  peptides 9
ACE inhibitory (ACEi) 7, 8, 15, 19, 27, 28
  activities 27
  activity *in vitro* and antihypertensive
   activity 28
  properties 28
Acetylcholinesterase 90
AChE 90
  inhibition and antioxidant activity 90
  inhibition and antioxidant activity
   combination 90
Acids 6, 15, 30, 32, 45, 48, 62, 67, 111, 126,
   127, 136, 137, 139, 140, 143, 157, 159,
   165, 168, 169
  1,3,7-trimethyluric 139
  3-butanoic 126
  3-hydroxypilocarpic 140
  3-Pyridylacetic 126
  4-quinolone-3-carboxylic 48
  5-hydroxyindoleacetic 137, 143
  5-trimethoxybenzoic 136, 137
  7-dimethyl uric 139
  7-methyluric 139
  acetic 6, 32
  arachidonic 165
  arylcarboxylic 67
  betulinic 168, 169
  butanoic 126
  formic 32
  lauric 30
  methyluric 139
  nitrous 67
  oleanolic 157, 165
  oxolinic 45
  pipemidic 45
  propionic 32
  reserpic 136, 137
  salvianolic 159, 168
  trimethyluric 140, 143
  ursolic 157, 165
Activation 3, 78, 154, 155, 163, 164, 165, 169,
   172
  carcinogenic 163
  caspase-3 172
  metabolic 78
  of oncogenes 154, 155
Active 30, 131, 136
  endolytic cysteine protease 30
  human liver cytochrome enzyme 136
  microsomal enzyme 131
Active metabolites 127, 130, 131, 133, 136,
   137, 139, 144
  identified 127
  of dihydroquinidine 131
Activity 26, 27, 30, 44, 45, 46, 50, 51, 62, 63,
   69, 70, 78, 81, 83, 89, 90, 124, 142, 143,
   166, 168, 169, 170, 171
  α-amylase 31
  anticancer 170
  antimicrobial 26, 30, 78
  antimycobacterial 62
  antiplasmodial 81
  anti-thrombotic 30
  antitumor 89
  biodynamic 124
  hTERT 166
  motor 143
  promising 90
  spectrum 45
  therapeutic 171
Adenomatous polyposis coli (APC) 155
ADMET 156, 173

Differential scaning calorimetry 17
Digestive enzymes 24
Diseases 28, 29, 73, 84, 151, 152, 163, 170, 171
  cardiovascular 29, 84
  complex 151
  coronary heart 28
  curing 163
  infective 73
  renal 28
DLS 116, 121
  analysis 116
  measurement 121
DNA 45, 60, 69, 79, 153, 154, 155, 162, 163, 166, 169
  adduct formation 162
  cleaved 169
  enzyme-cleaved 169
  new telomeric 166
  repair mechanism 154, 155, 163
  replication 69, 169
  synthesis 45, 154
DNA gyrase and topoisomerase IV 43, 45, 47, 48, 69, 84
  inhibition of 48, 69

E

Effects 13, 162, 164, 169, 170, 171
  antagonistic 170
  antibacterial 13
  anti-cancer 164, 169
  anti-tumor 171
  enhancing 162
Efficacy 43, 45, 50, 69, 152, 162, 170, 173
  therapeutic 50, 170
Efflux pumps 44, 50, 51, 64, 85
  bacterial 64
  inhibitors (EPIs) 50, 51
  up-regulated 51
Efflux pump system 50, 51
  inhibitors 51
EGF-mediated cell migration 159, 169
Elemental analysis 19, 20, 21
Emulsifying 31
Endoglucanase 8, 11
Enzymatic hydrolysis 1, 3, 4, 9, 15, 29, 127
  microwave-assisted 15
Enzymatic 7, 12
  hydrolysis by alkaline protease 12

methods 7
Enzyme(s) 4, 5, 6, 7, 8, 13, 14, 15, 19, 26, 27, 28, 29, 31, 32, 45, 139, 151, 152, 156, 162
  activities 6
  activity, main 7
  -assisted extraction peanuts kernels 26
  cytochrome P450 139
  DNA 45
  liver 156
  induction 162
  inhibition 162
  specificity 4
Epidermal growth factor receptors (EGFR) 159, 161, 168, 169
ER-independent mechanisms 166
ER-positive breast cancer 166
Estrogen receptor 158, 166
  signaling 158, 166
Expression 155, 159, 162, 168, 169, 170, 172
  gene 162
  ligand-induced 159
  suppressing GFRs 168
Extracellular 4, 168
  domain 168
  proteases 4
Extraction 2, 6, 7, 8, 10, 15, 17, 18, 22, 26, 110, 127
  alkaline-ethanol 18
  aqueous 18
  by chemical methods and hydrolysis 6
  by enzymatic methods and hydrolysis 7
  by physical/physical-chemical methods and hydrolysis 8
  electro-assisted 8
  enzyme-assisted 22
  high pressure-assisted 8
  of phenolic compounds and carbohydrate hydrolysis 10
  oil 7
Extraction conditions 7, 15
  subcritical water 15
Extraction methods 6, 15, 23, 136
  liquid-liquid 136
Extracts 8, 151, 163, 164
  crude 164
  herbal 151

concept 64
technique 43
Molecules 50, 57, 59, 64, 86, 124, 144, 152,
    156, 162, 164, 165, 170, 172
  anti-apoptotic 165
  antibiotic 86
  cellular adhesion 165
  chemopreventive 162, 165
  complex 59
  convenient chief 57
  novel 64
  roscovitine 172
  signalling 156
  therapeutic 152
Monoclonal antibodies 168
Mono-hydroxylation product 134
Mucosal signaling molecule 143
Mucuna pruriens 137
Multi-angle light scattering (MALS) 108, 109,
    112, 113, 114
Mutation in GFR genes 155
Myc genes 155
Mycobacterium tuberculosis 92

N

Natural 30, 32, 54, 124, 151, 163, 166, 169
  additives 32
  bioactives 151, 166, 169
  bioactive substance 151
  drug discovery progam 163
  electrophilic 54
  medicines 124
  preservatives 30
Natural product 163, 164
  -based drugs 163
  research 164
Neurodegenerative disorders 84
Neurological disorders 165
Neutral protease 10, 14
Neutrase 4, 13, 14, 18, 19, 28, 31
  hydrolyzate 18
Nicotine 124, 125, 126, 142
Nitrogen-containing compounds 125
Norscopolamine 127, 129
  conjugated 129
Nuclear magnetic resonance 141
Nutraceuticals 3, 28, 31, 170, 171
  effective 170
  promising 170

O

Oil 2, 7, 8, 26, 116
  extract 26
  silicone 116
Olive 7, 13, 14, 16, 23, 28, 29
  seeds 7, 23, 28, 29
  stones 13, 14, 16
Oncogenes 153, 154, 155
  viral 155
Oncogenesis 170
ORAC activity 27
Oral mucositis 164
Oxidation 139, 140, 141
  process 139
  reactions 139
Oxidative stress 90, 165, 168
Oxygen radical absorbance capacity (ORAC)
    10, 11, 12, 15, 27

P

Palm kernel expeller proteins 13
Papain activity 6, 30
Parasympathetic nervous system 142
Pathogens 43, 44, 70, 71, 78, 87, 91, 152
  anaerobic 78
  gram-negative respiratory tract 70
  intracellular 87
  persistent 91
  resistant 43, 71
Pathways 108, 120, 121, 133, 151, 155, 156,
    165, 167, 168, 172
  caspaseand non-caspase-dependent 172
  mechanistic 167
  mitochondrial 167
  molecular 151
  regulated 167
  signalling 155
  targeting multiple 165
Pectinase 7, 12, 13
Penicillium citrinum 160
Pepsin 5, 14
Peptidases 1, 2, 3, 4, 5, 6, 7, 8, 15, 17, 18, 27,
    31
  alkaline 7
  dipeptidyl 5
  microbial 8